Reyolando M.L.R.F. Brasil,
Marcelo Araujo da Silva

Project
Optimization

Using MATLAB and SOLVER

DE GRUYTER

Authors
Prof. Reyolando M.L.R.F. Brasil
Federal University of ABC
Av. dos Estados 5001
09210-580 SANTO ANDRÉ-Sao Paulo
Brazil

Prof. Marcelo Araujo da Silva
Federal University of ABC
Av. dos Estados 5001
09210-580 SANTO ANDRÉ-Sao Paulo
Brazil

ISBN 978-3-11-062561-5
e-ISBN (PDF) 978-3-11-062562-2
e-ISBN (EPUB) 978-3-11-062567-7

Library of Congress Control Number: 2021939932

Bibliographic information published by the Deutsche Nationalbibliothek
The Deutsche Nationalbibliothek lists this publication in the Deutsche Nationalbibliografie;
detailed bibliographic data are available on the Internet at http://dnb.dnb.de.

© 2021 Walter de Gruyter GmbH, Berlin/Boston
Cover image: ©Araujo da Silva, Marcelo
Typesetting: Integra Software Services Pvt. Ltd.
Printing and binding: CPI books GmbH, Leck

www.degruyter.com

Reyolando M.L.R.F. Brasil, Marcelo Araujo da Silva
Project Optimization

Also of Interest

Nature-Inspired Optimization Algorithms.
Recent Advances in Natural Computing and Biomedical Applications
Aditya Khamparia, Ashish Khanna, Nhu Gia Nguyen, Bao Le Nguyen,
2021
ISBN 978-3-11-067606-8, e-ISBN 978-3-11-067611-2

Advanced Reactor Modeling with MATLAB.
Case Studies with Solved Examples
Riccardo Tesser, Vincenzo Russo, 2021
ISBN 978-3-11-063219-4, e-ISBN 978-3-11-063292-7

MATLAB Programming.
Mathematical Problem Solutions
Dingyü Xue, 2020
ISBN 978-3-11-066356-3, e-ISBN 978-3-11-066695-3

Data Science for Supply Chain Forecasting.
Supramolecular Inclusion in Solution
Nicolas Vandeput, 2021
ISBN 978-3-11-067110-0, e-ISBN 978-3-11-067112-4

Inventory Optimization.
Models and Simulations
Nicolas Vandeput, 2020
ISBN 978-3-11-067391-3, e-ISBN 978-3-11-067394-4

Preface

In all branches of human endeavor, in which it is desired to realize something, it is necessary to develop a project. This should define precisely the objective(s) to be achieved, all variables that affect its result, and the resources available.

It is a fact of life that resources are always limited and a project must, in principle, provide the best possible solution within these limitations. This process is called optimization.

Some privileged few have the ability to glimpse the best solution by intuition, heuristically. The vast majority of mortals need some tool that will guide them among the many possibilities in general. This tool is mathematics. In particular, differential calculus is a technique that deals with the measurement of changes in variables upon the value of a function. One of its virtues is to enable the determination of variables or values that maximize or minimize a function.

This book is intended to introduce students and practitioners from a wide range of project designers to tools to help them optimize their projects. The authors are structural engineers and therefore many of the examples will be of structural engineering, but the techniques available are common to various areas of knowledge and pervade disciplinary divisions.

The authors publicly express their debt of gratitude with some sources from which they borrowed heavily in this text, among them, the works of J. S. Arora.

We also warmly thank our former student, Aerospace Engineer Jean Hermes Carvalho Vasco, for his careful and competent rendition of the figures in this book, and our student of Aerospace Engineering Lucas Moura de Almeida, for his contribution on topology optimization, Chapter 8.

The eldest of the two authors thanks life for the opportunity, at this stage, to leave some contributions to the new generations and, moreover, to have what to leave them.

Cheers.
The authors

https://doi.org/10.1515/9783110625622-202

Contents

About notation

Attempts have been made to keep a single notation for various quantities discussed throughout this book. In some cases, small variations were accepted for consistency with norms, market usage, and original reference works.

In general, in this text, bold capital letters denote matrices, lowercase bold letters denote vectors, and italicized letters denote scalar magnitudes.

The letter T overwritten to the right of an array indicates its transpose, that is, permutation of rows by columns. An exponent −1 to the right of a matrix indicates its inverse.

Two vertical bars to the right and left of a matrix or vector denote a norm of the same; in the case of a scalar, they denote an absolute value.

Chapter 1
Fundamental ideas

1.1 Introduction

Optimization is the process of determining, among several options, an object that is the best possible individual of its kind within certain criteria of choice and limitations, such as available resources. In designing a venture, all the time, the best performance is sought in its various disciplines: analysis, design, manufacturing, sales, research, development, and so on. This is pretty much the definition of a PROJECT or a DESIGN.

The traditional design process is based on the analysis of several solutions and the feasibility of their execution. In this process, there is no formal way of improving a given project and the designer can only improve it based on his/her intuition and experience. With a certain solution at hand, a decision needs to be made: accept the project as final or refine it. It is clear that this method strongly depends on the intuition, experience, and skill of the designer.

On the other hand, the optimal design process is more structured. In this approach, the design variables are first identified, the objective function which measures the relative merit of a solution, and the existing design constraints, must be defined according to the design variables. Once these quantities are defined, an appropriate optimization method can be used to optimize an initially estimated design. The designer still needs to adopt an initial design, but project enhancement now depends not only on his/her experience but on an optimization algorithm. As a result, the optimum design process can lead to safer and more economical solutions, in a relatively short time, using a computerized process.

Consider the example shown in Figure 1.1, where a metal plate that is subjected to a specified loading (a concentrated horizontal load applied at its upper left-hand side) has a initial square shape, as shown in Figure 1.1(a). In this figure, the stress distribution, in kgf/cm^2, is also shown for the initial design. After the application of an optimization method developed by the authors, in Chapter 8, the final design is shown in Figure 1.1(b). Note that a good part of the plate mass, where the stresses were low in case (a), marked with black color, was eliminated from the plate domain in the optimization process. In this case, the optimization process reduced the plate mass by more than 50%, producing a significant reduction in the amount of material used to make the plate.

In mathematical terms, optimization tries to find extreme (maximum or minimum) values of a function (the objective function) that depends on one or more design variables, subject to equality or inequality constraints. It is a vast field of knowledge and research, applied to all areas of engineering, science in general, logistics, and so on. In business management, it is sometimes renamed operational research.

https://doi.org/10.1515/9783110625622-001

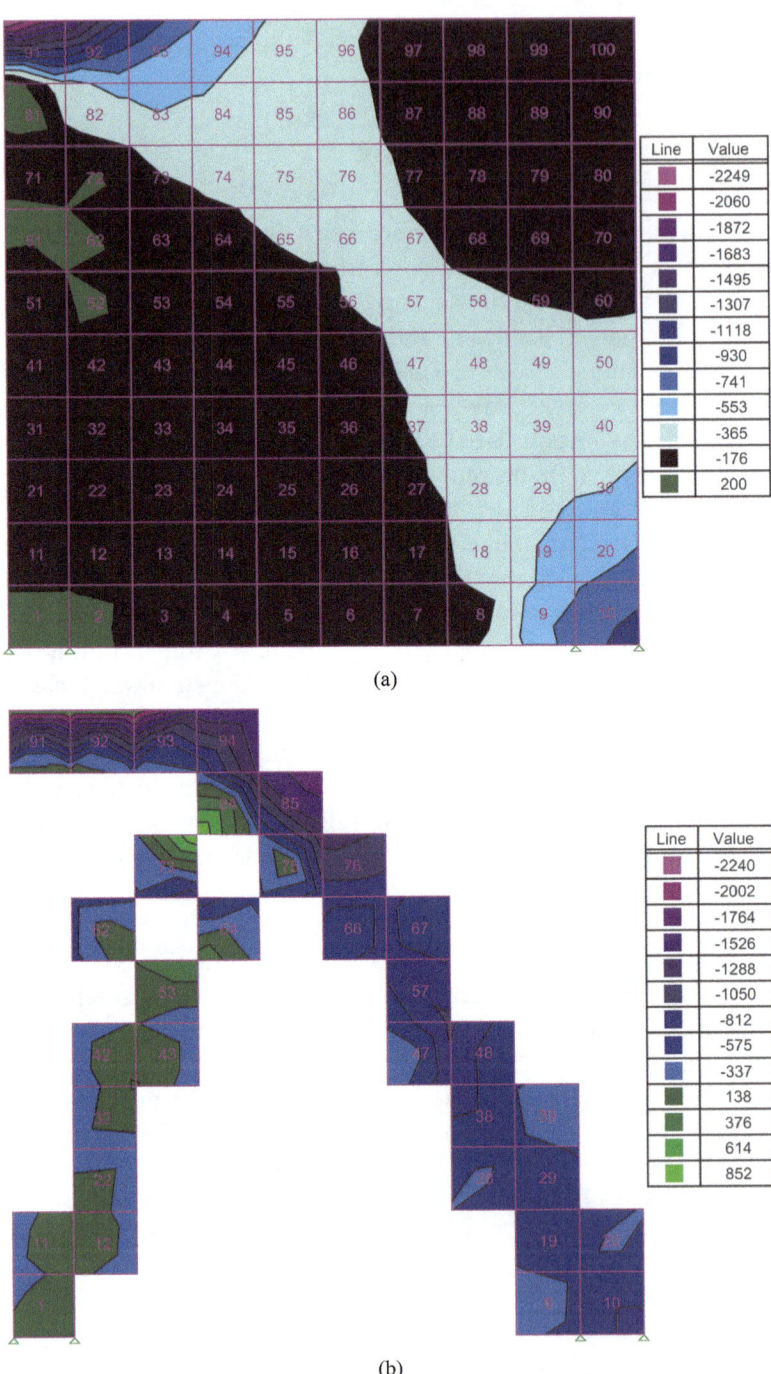

Figure 1.1: Example of shape optimization of a metal plate subjected to concentrated loading.

1.2 Elements of an optimization problem

1.2.1 Design variables

The design variables are a group of functions each expressing the value (variable during the optimization process) of a given parameter of a given project. Each design variable is independent of the others, being able to assume a certain value in a given continuous domain, or continuous in parts, or discrete. A beam of rectangular section, made of a certain material, destined to bridge a certain span and supporting a certain load, has two design variables: the width and height of its section. The vector of the design variables will be denoted here by the vector **x**. In the case in question, the vector can be written as $\mathbf{x} = [b_w\ h]^T$, as shown in Figure 1.2. The optimization problem can be defined as finding suitable values of b_w and h, such that the beam supports loading without rupture, excessive displacements, or cracks that may impair its long-term performance.

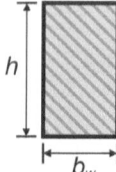

Figure 1.2: Rectangular transverse section of a beam.

The design of the cylindrical soda can shown in Figure 1.3, given the desired volume and the packing pressure, has three design variables: its diameter, its height, and the thickness of the sheet metal. The vector of the design variables in this case is $\mathbf{x} = [D\ h\ e]^T$. Depending on the formulation, other design variables may be adopted, such as different thicknesses for the lateral e_l, for the top e_t, and for the base e_b. Thus, we would have $\mathbf{x} = [D\ h\ e_l\ e_t\ e_b]^T$. It is clear that many formulations may exist for the same problem.

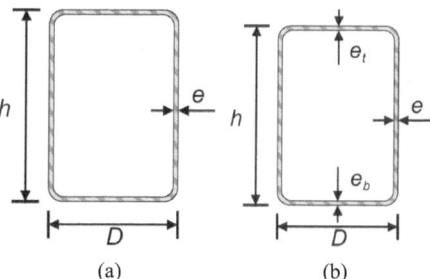

Figure 1.3: Section of a soda can displaying the design variables.

Design variables can be related to materials, topology, configuration, component capacity, and so on. Material-related design variables are used in selecting the type of materials adopted: steel, concrete, polymers, and others. They are discrete variables that represent the physical and mechanical properties of materials. Topological variables are introduced if the shape or layout of the system is being optimized. Variables of a component capacity can be from the productive capacity of certain equipment used in a production line up to the resistance of certain materials. The type of profile to be adopted in a design of metal structures can be considered either a configuration variable or a topological one.

Selecting design variables is an important step, since the whole formulation of the problem depends on their definition. They should be selected in such a way that the calculation process is implementable and the final design is practical. The feasible domain for solving a given problem usually increases in proportion to the increase in the number of design variables. In other words, increasing the number of design variables usually results in a better design. In this book, as already mentioned, the design variables will be represented by a vector $\mathbf{x}$ as follows:

$$\mathbf{x} = \begin{bmatrix} x_1 & x_2 & x_3 & \cdots & x_n \end{bmatrix}^T \tag{1.1}$$

where n is the total number of design variables.

In the case of discrete design variables, they must satisfy the condition:

$$x_i \in \mathbf{x}_i \equiv \left\{ x_{i1} \quad x_{i2} \quad \cdots \quad x_{iN_{Ei}} \right\}, \quad i = 1, \ldots, n \tag{1.2}$$

and $x_{i1}, x_{i2}, \ldots, x_{iN_{Ei}}$ are the N_{Ei} possible discrete values of variable x_i. For example, a steel rebar CA-50 (yield stress 500 MPa) is available in the following diameter schedule {6.3 8 10 12.5 16 20 25 32} mm. This also occurs with the gauge of electrical cables that have predefined dimensions.

1.2.2 Objective function or cost function

The objective function, or cost function, determines the relative merit of several designs for a given system. The selection of the objective function is an important task because the projects are improved by minimizing or maximizing its value.

Considering an example of civil structures, in most of the structural optimization problems, the weight of the structure is chosen as an objective function. This fact is due to the great ease of computation of this quantity and also because its value is directly related to the cost of the materials used. More efficient use of materials will minimize the cost of construction when all other factors, such as manufacturing, transportation, assembly, and maintenance costs, remain constant. In structural optimization, these factors are usually not constant, but rather functions of the design variables. For example, the cost of transporting a particular prefabricated structural element depends on its weight and dimensions.

There are other costs behind a construction cost that must be considered in the sizing processes. These can be the time of construction, costs related to the ruin of the structure, the efficiency of the structure, among others. The cost related to construction time can be easily computed, while the cost of ruin may in some cases be impossible to determine. The ruin of a structural system is intrinsically related to the safety adopted both in the sizing process and in the construction process. The appearance of an ultimate or service limit state in the structure may be due to the combination of several random factors arising from the following causes: (a) uncertainties regarding values considered as resistances of the materials used, taking into account not only the conditions of execution and control of the work, as well as some parameters that affect the limit state in question (such as long duration load and fatigue); (b) errors made regarding the geometry of the structure and its sections; (c) inaccurate assessment of direct, indirect, or exceptional actions, due to the impossibility of defining them, a priori, with absolute precision, throughout the useful life of the structure; (d) divergence between the calculated values and the actual values of the internal forces, due to simplifying assumptions usually used in the calculation. A good objective to be sought in the dimensioning of a structure is that of minimum cost, keeping forces below a previously established value of the probability of the appearance of a limit state. The purpose of applying, in this design, the principles of probabilistic theory would be to obtain, with appropriate security, the optimal cost of the structure. This should consider, among the various factors, moral and psychological considerations (which are difficult to quantify), as well as the value of human life and the reaction of public opinion to the occurrence of some accident.

Before trying to formulate all the factors involved in a sizing process, it is important to know if they actually influence the solution. It would not be desirable to consider a too general objective function, because the result can be a flat objective function that is not sensitive to changes in design variables and does not result in an improvement of the design initially adopted. Since the most important factors in cost computing are determined, they can be calculated according to their related design variables.

Sometimes it is desirable to minimize or maximize several objective functions simultaneously. This is called multicriteria optimization or multipurpose optimization. This type of problem can be defined as determining a vector of design variables that satisfies the constraints and optimizes a vector function whose components are the various objective functions. Objective functions considered in this type of problem may generally be in conflict with one another. As an example, in the simultaneous optimization of a structure with an incorporated vibration control system, both the minimization of the cost of the structure and the minimization of the oscillations should be treated as objective functions. In this case, it is clear that minimizing the cost of the structure would mean reducing the dimensions of the sections of the structural elements, which would lead to an increase in displacements.

A general objective function for a dynamic system (variable in time) can be defined as

$$f(\mathbf{x}, T) = \bar{f}(\mathbf{x}, T) + \int_0^T \tilde{f}(\mathbf{x}, \mathbf{z}, \dot{\mathbf{z}}, \ddot{\mathbf{z}}, t) dt \tag{1.3}$$

where $\mathbf{z}$ is the vector of state variables such as displacements, electric charge, temperature, and T is the considered design time interval. It is assumed that the objective function is continuous and differentiable. State variables are considered as continuous functions of time and are determined by the integration of the state equations of the system. In the case of electrical engineering, a state equation is given, for example, by associations of electric circuits, whereas in structural engineering it is given by the equation of motion. Equation (1.3) can represent any cost function. For example, $\bar{f}$ can represent the mass of the structure, or the length of the electric cables, while $\tilde{f}$ can represent displacement, or any other function involving the state variables.

1.2.3 Design constraints

For the problems described and resolved in this book, design constraints are divided into two groups: static constraints and dynamic constraints. Dynamic constraints are imposed over the entire design time interval $t \in [0,T]$ in which the system is analyzed. Limits for the values assumed by stresses, displacements, and accelerations are examples of this type of variables. On the other hand, the static constraints are independent of time and are related to the geometric limits of the structure, established intervals for the natural frequencies of vibration, static displacements, static stresses on the ground, and physical limits for the design variables. Limits for potential losses in the case of electrical projects are imposed constraints.

A general way to represent the static constraints is

$$g_i = \bar{g}_i(\mathbf{x}, T) + \int_0^T \tilde{g}_i(\mathbf{x}, \mathbf{z}, \dot{\mathbf{z}}, \ddot{\mathbf{z}}, t) dt \begin{cases} = 0 \text{ for } i = 1, ..., l \\ \leq 0 \text{ for } i = l+1, ..., m \end{cases} \tag{1.4}$$

and a general form of dynamic constraints is:

$$g_i = \tilde{g}_i(\mathbf{x}, \mathbf{z}, \dot{\mathbf{z}}, \ddot{\mathbf{z}}, t) \begin{cases} = 0 \text{ for } i = m+1, ..., l' \\ \leq 0 \text{ for } i = l'+1, ..., m' \end{cases}, \text{ for } t \in [0, T] \tag{1.5}$$

1.3 The standard optimization problem

In general, an optimization process can be described according to the flowchart shown in Figure 1.4.

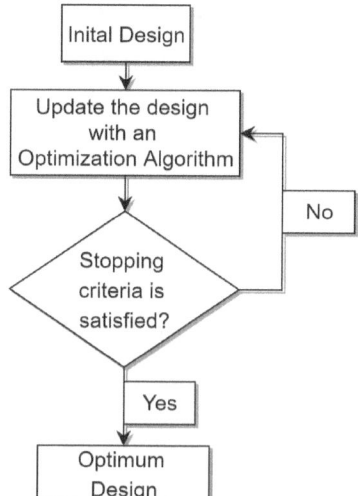

Figure 1.4: Generic flowchart of an optimization problem.

It should be noted, in Figure 1.4, that the classical optimization process starts from an initial design or a group of initial design that is improved according to a particular optimization method.

A general optimization problem can be formulated in the standard form that follows.

Let a given problem be defined by the values of a vector of n **design variables**

$$\mathbf{x} = \begin{bmatrix} x_1 & x_2 & \cdots & x_n \end{bmatrix}^T \tag{1.6}$$

Determine $\mathbf{x} \in \Re^n$ that minimizes the **objective function** $f(\mathbf{x})$, subjected to **equality constraints**

$$h_j(\mathbf{x}) = 0, \quad j = 1, \ldots, p \tag{1.7}$$

and **inequality constraints**

$$g_i(\mathbf{x}) \leq 0, \quad i = 1, \ldots, m \tag{1.8}$$

Functions f, g, and h, may, in general, be nonlinear.

If the maximum value of function $f(x)$ is desired, instead of a minimum, it suffices to find the minimum of the function with inversed sign, $F(x) = -f(x)$.

For low-dimensional problems, the problem can be solved by simply inspecting graphs of the functions f, g, and h, as in Chapter 3.

1.4 Examples

E1. An example of operational research, based on Arora (2016), is the profit maximization of a company that manufactures two types of toys, A and B. Using the available resources, either 28 A toys or 14 B toys can be produced per day. The sales department can sell either 14 A toys or 24 B toys. The shipping department cannot handle more than 16 toys per day. The company profits ∈ 400 per toy A and ∈ 600 per toy B. How many toys of each type give maximum profit?

Design variables:

x_1 = number of toys A, x_2 = number of toys B

Objective function (profit), to be maximized: $F(\mathbf{x}) = 400x_1 + 600x_2$

Inequality constraints:

$$x_1 + x_2 \leq 16 \quad \Rightarrow \quad g_1(\mathbf{x}) = x_1 + x_2 - 16 \leq 0 \qquad \text{(shipping)}$$

$$x_1/28 + x_2/14 \leq 1 \quad \Rightarrow \quad g_2(\mathbf{x}) = x_1/28 + x_2/14 - 1 \leq 0 \qquad \text{(production)}$$

$$x_1/14 + x_2/24 \leq 1 \quad \Rightarrow \quad g_3(\mathbf{x}) = x_1/14 + x_2/24 - 1 \leq 0 \qquad \text{(sales)}$$

E2. Consider the prismatic cantilever beam shown in Figure 1.5.

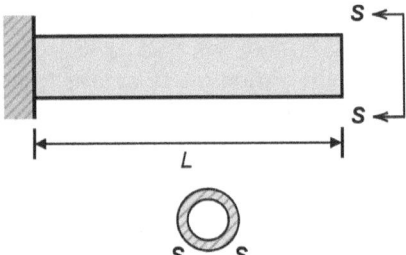

Figure 1.5: Cantilever beam.

The transverse section is a circular ring. The external diameter is D and the wall thickness e. The beam stiffness is $k = 3EI/L^3$, where E is the elastic modulus, I is the moment of inertia of the cross section, and L is the length of the beam. A lumped mass at the tip of the beam equivalent to its self-weight is M. A good approximation of M is about one-fourth of the total mass of the beam, the product of its total volume times the mass per unit volume ρ. The first vibration frequency is

$$f_1 = \frac{1}{2\pi} \sqrt{\frac{k}{M}}. \tag{1.9}$$

A classic optimization problem is to minimize the mass of the beam by imposing that the first frequency of vibration must exceed a certain minimum value. The design variables are defined as

$x_1 = D$ (external diameter of the section), $x_2 = e$ (wall thickness)

Objective function (mass), to be minimized: $f(\mathbf{x}) = \rho \dfrac{\pi}{4}\left[x_1{}^2 - (x_1 - 2x_2)^2\right]L$

Inequality constraints:

$f_1 \quad \geq f_{\min}$ (minimum value of the first natural vibration frequency)

$D_{\min} \leq x_1 \leq D_{\max}$ (lower and upper values of the external diameter)

$e_{\min} \leq x_2 \leq \dfrac{x_1}{2}$ (lower and upper values of the thickness)

This type of problem is quite common in telecommunication towers, where it is desirable to have $f_{\min}$ of at least 1 Hz, or in wind power towers where the $f_{\min}$ should be larger than 0.5 Hz. In a real problem, other conditions must be considered, such as applied loads, the resistance of the materials used, limits for static and dynamic displacements, fatigue, and so on.

Chapter 2
Essential mathematical optimization tools

In most of the projects, it is not straightforward to determine their best solution. Some privileged few have the ability to glimpse the best solution by intuition, heuristically. The vast majority of designers need some tool to guide them among the many possibilities, in general existing. This tool is mathematics. In particular, differential calculus is a technique that deals with the measurement of the effect of variation of the variables on the value of a function. One of its virtues is to enable the determination of variable values that maximize or minimize a function. The variational calculus is also widely used, especially in dynamic problems and also in energy methods.

Differential calculus was the great legacy of two of the greatest geniuses that mankind has ever produced, contemporaries Newton and Leibnitz. As for the variational calculus, one can distinguish the great contributions of Lagrange and Hamilton.

This chapter deals specifically with the mathematical concepts involved in optimization processes. Although some of the contents may be familiar to those who have not had their mathematical education neglected, it is not advisable to skip this chapter altogether and go forward to applications, coming back in case of doubts in these mathematical tools.

2.1 Vectors and matrices

A vector $\mathbf{x}$, or a **point** in related space, is, in a simplistic way, an ordered set of n values x_i ($i = 1, \ldots, n$), usually represented in a single column, which expresses the state of a system. Thus, the position of a point in the three-dimensional space in which we live can be expressed without ambiguity by three coordinates,

$$\mathbf{x} = \left\{ \begin{array}{c} x_1 \\ x_2 \\ x_3 \end{array} \right\} \tag{2.1}$$

If these coordinates vary in time, that is, if the particle is in motion, these coordinates are functions of time, not constants, and express a **trajectory** in that space. In this case, it is common to refer to the vector as a **field**.

In this text, vectors will be represented by lowercase Latin letters in bold.

The scalar product of two vectors is defined as

$$(\mathbf{x} \bullet \mathbf{y}) = \mathbf{x}^{\mathsf{T}} \mathbf{y} = \sum_{i=1}^{n} x_i y_i \tag{2.2}$$

https://doi.org/10.1515/9783110625622-002

In the so-called Einstein convention, the repetition of indexes in a monomial implies a summation for all the values of this index. Thus, Eq. (2.2) is, simply,

$$(\mathbf{x} \bullet \mathbf{y}) = x_i y_i \tag{2.3}$$

The norm or length of a vector is given by

$$\|\mathbf{x}\| = \sqrt{x_i x_i} = \sqrt{\mathbf{x} \bullet \mathbf{x}} \tag{2.4}$$

An array, or matrix, is an ordered set of values with a certain number n of rows and m of columns, and is represented here by bold Latin uppercase letters.

2.2 Functions and their derivatives

2.2.1 Single variable functions

A function of one variable $f = f(x)$ is a procedure that transforms a numeric value of x into another numeric value of f.

In a single variable function, without constraints, the determination of maxima and minima is a classical problem of differential calculus, and readers are referred to the standard texts of that discipline. Note that the function must be necessarily nonlinear, since the quest for maximum and minimum of a linear function, without constraints, is meaningless.

2.2.2 Functions of several variables and vector functions

A function $f = f(\mathbf{x})$, or

$$f(\mathbf{x}) = f(x_1, x_2, \ldots, x_n) \tag{2.5}$$

is a procedure that transforms the values of a vector or point $\mathbf{x}$ into a number f.

In the same way, one can have a vector of m functions of n variables in the form

$$\mathbf{g}(\mathbf{x}) = \left\{ \begin{array}{c} g_1(\mathbf{x}) \\ g_2(\mathbf{x}) \\ \vdots \\ g_m(\mathbf{x}) \end{array} \right\} \text{ where } g_j(\mathbf{x}) = g_i(x_1, x_2, \ldots, x_n), \quad j = 1, \ldots, m \tag{2.6}$$

In the search of maximum and minimum of functions of these types arises the necessity of the calculation of partial derivatives. The first-order partial derivatives of (2.5) are

$$\frac{\partial f(\mathbf{x})}{\partial x_i}; \quad i = 1, \dots, n \tag{2.7}$$

which can be arranged in a column vector called the gradient vector of the function:

$$\mathbf{c}(\mathbf{x}) = \vec{\nabla} f(\mathbf{x}) = \frac{\partial f(\mathbf{x})}{\partial x_i} = \left\{ \begin{array}{c} \frac{\partial f(\mathbf{x})}{\partial x_1} \\ \frac{\partial f(\mathbf{x})}{\partial x_2} \\ \vdots \\ \frac{\partial f(\mathbf{x})}{\partial x_n} \end{array} \right\} \tag{2.8}$$

where the **nabla** operator $\vec{\nabla}$ has been used.

Each component of Eq. (2.8) can be differentiated again with respect to a variable to obtain second-order partial derivatives

$$\frac{\partial^2 f(\mathbf{x})}{\partial x_i \partial x_j}; \quad i, j = 1, \dots, n \tag{2.9}$$

Equation (2.9) may be arranged in an n^2 array, called the Hessian matrix:

$$\mathbf{H}(\mathbf{x}) = \vec{\nabla}\vec{\nabla}^T f(\mathbf{x}) = \begin{bmatrix} \frac{\partial^2 f(\mathbf{x})}{\partial x_1 \partial x_1} & \frac{\partial^2 f(\mathbf{x})}{\partial x_1 \partial x_2} & \cdots & \frac{\partial^2 f(\mathbf{x})}{\partial x_1 \partial x_n} \\ \frac{\partial^2 f(\mathbf{x})}{\partial x_2 \partial x_1} & \frac{\partial^2 f(\mathbf{x})}{\partial x_2 \partial x_2} & \cdots & \frac{\partial^2 f(\mathbf{x})}{\partial x_2 \partial x_n} \\ \vdots & \vdots & & \vdots \\ \frac{\partial^2 f(\mathbf{x})}{\partial x_n \partial x_1} & \frac{\partial^2 f(\mathbf{x})}{\partial x_n \partial x_2} & \cdots & \frac{\partial^2 f(\mathbf{x})}{\partial x_n \partial x_n} \end{bmatrix} \tag{2.10}$$

There are cases where it is necessary to differentiate a vector of m functions of n variables as (2.6), with respect to the variables themselves. In this case, there is the so-called Jacobian matrix:

$$\mathbf{J}(\mathbf{x}) = \vec{\nabla}\mathbf{g}(\mathbf{x})^T = \begin{bmatrix} \frac{\partial g_1(\mathbf{x})}{\partial x_1} & \frac{\partial g_1(\mathbf{x})}{\partial x_2} & \cdots & \frac{\partial g_1(\mathbf{x})}{\partial x_n} \\ \frac{\partial g_2(\mathbf{x})}{\partial x_1} & \frac{\partial g_2(\mathbf{x})}{\partial x_2} & \cdots & \frac{\partial g_2(\mathbf{x})}{\partial x_n} \\ \vdots & \vdots & & \vdots \\ \frac{\partial g_m(\mathbf{x})}{\partial x_1} & \frac{\partial g_m(\mathbf{x})}{\partial x_2} & \cdots & \frac{\partial g_m(\mathbf{x})}{\partial x_n} \end{bmatrix} \tag{2.11}$$

Note that the Jacobian matrix is m x n, that is, it is not necessarily square!

2.3 Taylor series expansion

A basic tool in this book is the Taylor series, due to the British mathematician Brook Taylor (1685–1731). In the case of a single (scalar) function of a single variable, if its value and its derivatives are known at a certain point a, we can approximate the value of the function at another nearby point x, defined by the "distance" $d = x - a$, by a series expansion:

$$f(x) = \frac{1}{0!}d^0 f(a) + \frac{1}{1!}d^1 f'(a) + \frac{1}{2!}d^2 f''(a) + \frac{1}{3!}d^3 f'''(a) + \cdots \qquad (2.12)$$

where the primes to the right and above the function denote successive derivatives in x, calculated at point a.

In the case of a single function (scalar) of several variables (a vector of n variables), if its value is known at point $\mathbf{a}$ ($n \times 1$), we can approximate the value of the function at another nearby point $\mathbf{x}$ ($n \times 1$), defined by a "distance vector" $\mathbf{d} = \mathbf{x} - \mathbf{a}$ ($n \times 1$), by the series expansion:

$$f(\mathbf{x}) = f(\mathbf{a}) + \mathbf{c}(\mathbf{a})^T \mathbf{d} + \frac{1}{2!}\mathbf{d}^T \mathbf{H}(\mathbf{a})\mathbf{d} + \cdots \qquad (2.13)$$

where the transpose of the *gradient vector* of Eq. (2.8) and the Hessian matrix ($n \times n$) of Eq. (2.10) are used.

Finally, in the case of a vector of n functions of several variables (a vector of n variables, $n \times 1$), if its value at point $\mathbf{a}$ ($n \times 1$) is known, then one can approximate the value of these functions in another nearby point $\mathbf{x}$ ($n \times 1$), defined by a "distance vector" $\mathbf{d} = \mathbf{x} - \mathbf{a}$ ($n \times 1$), by the series expansion:

$$\mathbf{f}(\mathbf{x}) = \mathbf{f}(\mathbf{a}) + \mathbf{J}(\mathbf{a})\mathbf{d} + \cdots \qquad (2.14)$$

where the Jacobian matrix of Eq. (2.11) is used (necessarily $n \times n$ in this case).

For the sake of information, when an expansion is made about the origin, that is, $\mathbf{a} = \mathbf{0}$, the Taylor series is called the MacLaurin series, due to the Scotsman Colin MacLaurin (1698–1746).

2.4 Quadratic forms and defined matrices

A quadratic form is a special form of nonlinear function with only second-order terms, in general form

$$F(\mathbf{x}) = \sum_{i=1}^{n} \sum_{j=1}^{n} p_{ij} x_i x_j = \mathbf{x}^T \mathbf{P} \mathbf{x} \qquad (2.15)$$

where **P** is the matrix of the quadratic form. There are infinite such matrices, all asymmetrical, except a symmetric one given by

$$\mathbf{A} = \frac{1}{2}\left(\mathbf{P} + \mathbf{P}^{\mathbf{T}}\right) \tag{2.16}$$

Since **A** is symmetric, it is known, from linear algebra, that all its eigenvalues are real.

A quadratic form $F(\mathbf{x}) = \mathbf{x}^{\mathbf{T}}\mathbf{A}\mathbf{x}$ can be positive, negative, or zero for any **x**, and its matrix is said positive definite, positive semidefinite, negative definite, negative semidefinite, or indefinite, as follows:
1. Positive definite: $F(\mathbf{x}) > 0$ for every $\mathbf{x} \neq 0$; $\lambda_i > 0$.
2. Positive semidefinite: $F(\mathbf{x}) \geq 0$ for every $\mathbf{x} \neq 0$; $\lambda_i \geq 0$ (at least a zero value).
3. Negative definite: $F(\mathbf{x}) < 0$ for every $\mathbf{x} \neq 0$; $\lambda_i < 0$.
4. Negative semidefinite: $F(\mathbf{x}) \leq 0$ for every $\mathbf{x} \neq 0$; $\lambda_i \leq 0$ (at least a zero value).
5. Indefinite: if positive for some values of **x** and $\lambda_i > 0$, and negative for others and $\lambda_i < 0$.

2.5 Minima and maxima of functions

2.5.1 Optimization without constraints

Consider the problem of minimizing the function $f(\mathbf{x})$, $\mathbf{x} \in \mathbf{R}^n$. The problem may be set as

$$\text{minimize} f(\mathbf{x}) \tag{2.17}$$

$$\mathbf{x} \in \mathbf{R}^n.$$

We consider $f(\mathbf{x}) \in C_2$. Such a function $f(\mathbf{x})$ is called an objective function.

In this section, conditions will be established that must be satisfied by a point, so that it is a local minimum of problem (2.17). The convexity properties of the objective function will also be described, which ensure that the found point is also a global minimum point.

For a point $\mathbf{x}^\star$ to be a local minimum of problem (2.17), it is sufficient that the gradient of the objective function in $\mathbf{x}^\star$ be zero, that is, $\nabla f(\mathbf{x}^\star) = 0$ and the Hessian matrix $\nabla^2 f(\mathbf{x}^\star)$ to be positive definite, that is

$$\mathbf{d}^T \nabla^2 f(x^\star)\mathbf{d} > 0 \quad \forall \mathbf{d} \neq 0 \tag{2.18}$$

Let $f(\mathbf{x})$ be a convex function in $\mathbf{R}^n$ and Ω be the set of points $\mathbf{x} \in \mathbf{R}^n$ where $f(\mathbf{x})$ is a minimum. Then Ω is convex and every local minimum is a global minimum point.

Let $f(\mathbf{x}) \in C_1$ be a convex function. If there is $\mathbf{x}^\star \in \mathbf{R}^n$, then every $\mathbf{y} \in R^n$.

An explanation of these minimum criteria for an unconstrained function is appropriate here. A minimum point is one in which the value of a function can only grow at points in a small neighborhood. To determine these values, the Taylor expansion of Eq. (2.13) can be used. Transferring the first term from the right side to the left, the difference in the value of the function is obtained between the minimum point and a neighboring point

$$\Delta f = \mathbf{c}(\mathbf{x}^*)^T \mathbf{d} + \frac{1}{2!} \mathbf{d}^T \mathbf{H}(\mathbf{x}^*) \mathbf{d} + \cdots$$

Since the distance $\mathbf{d}$ is small, the first term of this expression, related to the first derivatives of the function, is dominant over the second. This distance $\mathbf{d}$ can be positive or negative, but this term must be null if the point is a minimum. As a result, the gradient at that point has to be zero. For a single variable, simply the first derivative must be null at a point of minimum.

It remains to examine the second term. This must be positive so that the difference obtained always grows around the point that is assumed to be a minimum, for any small distance $\mathbf{d}$. This implies that the Hessian matrix, of the second derivatives, must be positively defined. For a single variable, simply the second derivative must be positive at a point of minimum:

$$\nabla f(\mathbf{x}^*)^T (\mathbf{y} - \mathbf{x}^*) \geq 0 \tag{2.19}$$

then $\mathbf{x}^*$ is a global minimum point of $f(\mathbf{x})$.

Example with a single design variable: Consider radius R and height H to be the design variables for fixed volume V of fluid in a cylindrical vessel made of sheet metal with fixed thickness. The consumption of sheet metal, in area, is the objective function, or cost function,

$$A = \pi(R^2 + RH)$$

with the constant volume constraint

$$V = \pi R^2 H$$

These two equations can be joined into an equivalent objective function of a single variable, R:

$$f = R^2 + \frac{V}{\pi R}$$

The first-order necessary optimum condition

$$f' = 2R - \frac{V}{\pi R^2} = 0$$

corresponds to design variable values

$$R^\star = \sqrt[3]{\frac{V}{2\pi}} \text{ and } H^\star = \sqrt[3]{\frac{4V}{\pi}}$$

as possible minimum points. The verification by the sufficient condition is given by the curvature

$$f'' = 2 + \frac{2V}{\pi R^3} = 6$$

which is positive, indicating a minimum point.

2.5.2 Optimization with constraints

Consider the general optimization problem

$$\text{minimize} \quad f(\mathbf{x}), \quad \mathbf{x} \in \mathbf{R}^n$$

$$\text{subject to} \quad \begin{aligned} g_i(\mathbf{x}) &= 0, & i \in E \\ g_i(\mathbf{x}) &\leq 0, & i \in I \end{aligned} \tag{2.20}$$

where function $f(\mathbf{x})$ is the objective function, and functions $g_i(\mathbf{x})$ are called constraints. E is the set of equality constraint indexes and I is the set of inequality constraint indexes. The solution of Eq. (2.20) is called simply the solution, or the optimum point, denoted as $\mathbf{x}^\star$.

When a point $\mathbf{x} \in \mathbf{R}^n$ satisfies all constraints, it is said a viable point, and the set of all viable points is said as the viable region Γ.

We admit $f(\mathbf{x})$ and $g_i(\mathbf{x}) \in C_2[\mathbf{R}^n]$.

Constraints $\{g_i(\mathbf{x}), i \in I | g_i(\mathbf{x}) = 0\}$ are called **active constraints** in $\mathbf{x}$. We will indicate $I^\star$ as the set of the indexes associated with the active constraints in $\mathbf{x}$.

Constraints $\{g_i(\mathbf{x})| \ ||g_i(\mathbf{x})|| \leq \varepsilon\}$ are called **ε-active constraints** in $\mathbf{x}$.

We say that a point $\mathbf{x}_v$ that satisfies constraints $g_i(\mathbf{x})$, $i \in E \cup I^\star$, is a regular point if gradient vectors $\{\nabla g_i(\mathbf{x}_v), i \in E \cup I^\star\}$ are linearly independent.

If only equality constraint functions are considered, then the Lagrangian function is defined as

$$L(\mathbf{x}, \mathbf{v}) = f(x) + \sum_{j=1}^{p} v_j h_j(x) \tag{2.21}$$

where v_i, $j = 1, \ldots, p$ are the Lagrange multipliers.

We indicate $\nabla L(\mathbf{x}, \mathbf{v}) = \begin{bmatrix} \nabla_x L(\mathbf{x}, \mathbf{v}) \\ \nabla_v L(\mathbf{x}, \mathbf{v}) \end{bmatrix}$ as the gradient vector $L(\mathbf{x},\mathbf{v})$, where ∇_x indicates partial derivatives with respect to x_i, $i = 1,\ldots,n$, and ∇_v indicates derivatives with respect to v_j, $j = 1, \ldots, p$

The gradient vector of the Lagrangian function with respect to $\mathbf{x}$ in $(\mathbf{x}^\star,\mathbf{v}^\star) \in \mathbf{R}^{n+m}$ is

$$\nabla_x L\left(\mathbf{x}^*, \mathbf{v}^*\right) = \nabla f\left(\mathbf{x}^*\right) + \sum_{j=1}^{p} v_j^* \nabla h_j\left(x^*\right) \tag{2.22}$$

where $\mathbf{v}^\star$ is the vector of Lagrange multipliers at the optimum point.

Analogously, the Hessian of the Lagrangian function with respect to $\mathbf{x}$ at point $(\mathbf{x}^\star,\mathbf{v}^\star) \in \mathbf{R}^{n+p}$ is

$$\nabla_x^2 L\left(\mathbf{x}^*, \mathbf{v}^*\right) = \nabla f\left(\mathbf{x}^*\right) + \sum_{j=1}^{p} v_j^* \nabla h_j\left(x^*\right) \tag{2.23}$$

Example with two design variables: Consider radius R and height H to be the design variables for a cylindrical fluid vessel having fixed volume V in a vessel made of sheet metal with constant thickness. The consumption of sheet metal, in area, is the objective function, or cost function,

$$A = \pi(R^2 + RH)$$

with fixed volume constraint

$$h = \pi R^2 H - V = 0.$$

The Lagrangian is

$L = \pi(R^2 + RH) + u(\pi R^2 H - V)$, where u is the Lagrange multiplier.

To be stationary:

$$\frac{\partial L}{\partial R} = 2R + H + 2\pi v R H = 0$$

$$\frac{\partial L}{\partial H} = R + \pi v R^2 = 0$$

$$\frac{\partial L}{\partial u} = \pi R^2 H - V = 0$$

To get the problem solution:

$$R^* = \sqrt[3]{\frac{V}{2\pi}}; \quad H^* = \sqrt[3]{\frac{4V}{\pi}}; \quad u^* = -\frac{1}{\pi R} = \sqrt[3]{\frac{2}{\pi^2 V}}$$

2.5.3 The Karush–Kuhn–Tucker (KKT) conditions: equality constraints

Let $\mathbf{x}^\star$ be a local minimum point of problem (2.20). If $\mathbf{x}^\star$ is a regular point, then there are Lagrange multipliers $\mathbf{v}^\star$, so that $\mathbf{x}^\star$ and $\mathbf{v}^\star$ satisfy the following set of equations:

$$\nabla_x L\left(\mathbf{x}^*, \mathbf{v}^*\right) = \nabla f\left(\mathbf{x}^*\right) + \sum_{j=1}^{p} v_j^* \nabla h_j\left(x^*\right)$$

$$h_j\left(\mathbf{x}^*\right) = 0, \quad i \in E$$

$$v_j^* > 0, \quad i \in E \tag{2.24}$$

Equations (2.24) are called the Karush–Kuhn–Tucker (KKT) conditions.

For a point $\mathbf{x}^\star$ to be a local minimum of problem (2.20) it must be a regular point, satisfy the KKT conditions and

$$\boldsymbol{d}^T \nabla_x^2 L\left(\boldsymbol{x}^*, \boldsymbol{u}^*\right) \boldsymbol{d} \geq 0 \quad \forall \mathbf{d} \tag{2.25}$$

If the objective function and the constraints are convex, the problem is said to be of convex programming. For convex programming problems, the following results are valid:
–   Every solution $\mathbf{x}^\star$ of a convex programming problem is a global solution and the set of global solutions S is a convex set.
–   If in the convex programming problem, the objective function is strictly convex in $\mathbf{R}^n$, then every global solution is unique.
–   If in a convex programming problem, the functions $f(\mathbf{x})$ and $g_i(\mathbf{x})$ are continuous with continuous partial derivatives until the first order, and if the conditions of KKT are satisfied in $\mathbf{x}^\star$, then point $\mathbf{x}^\star$ is a global solution to the convex programming problem.

2.5.4 Problems with general constraints using slack variables

Consider the general problem of optimization of several variables $\mathbf{x}$, with objective function $f(\mathbf{x})$, subject to p equality constraint functions $h(\mathbf{x})_j = 0, j = 1 \text{ a } p$ and m inequality constraint functions $g_i(\mathbf{x}) \leq 0, i = 1 \text{ a } m$, which can be written in the form of a vector as follows:

$$\mathbf{g}(\boldsymbol{x}) = \left[g_1(\boldsymbol{x}) \quad g_2(\boldsymbol{x}) \quad \cdots \quad g_m(\boldsymbol{x})\right]^T \tag{2.26}$$

The latter can be transformed into equality constraints by the addition of *slack variables* s_i, so that

$$g_i(\mathbf{x}) + s_i^2 = 0 \tag{2.27}$$

rendering vector

$$\mathbf{s} = \begin{bmatrix} s_1 & s_2 & \dots & s_m \end{bmatrix}^T \tag{2.28}$$

For these new relations of equality, Lagrange multipliers are adopted:

$$u_i \geq 0 \tag{2.29}$$

rendering vector

$$\mathbf{u} = \begin{bmatrix} u_1 & u_2 & \dots & u_m \end{bmatrix}^T \tag{2.30}$$

We now write a scalar **Lagrangian function** L, or Lagrangian, as

$$L(\mathbf{x}, \mathbf{u}, \mathbf{v}, \mathbf{s}) = f(\mathbf{x}) + \mathbf{v}^T \mathbf{h}(\mathbf{x}) \, \mathbf{u}^T \left(\mathbf{g}(\mathbf{x}) + \mathbf{s}^2 \right) \tag{2.31}$$

Using KKT, it is possible to set a straightforward procedure for the optimization general problem as follows:
1. Write the Lagrangian

$$L(\mathbf{x}, \mathbf{v}, \mathbf{u}, \mathbf{s}) = f(x) + \sum_{j=1}^{p} v_j h_j(\mathbf{x}) + \sum_{i=1}^{m} u_i (g_i(\mathbf{x}) + s_i^2) \tag{2.32}$$

2. Compute the gradient conditions

$$\frac{\partial L}{\partial x_k} = \frac{\partial f}{\partial x_k} + \sum_{j=1}^{p} v_j^* \frac{\partial h_j}{\partial x_k} + \sum_{i=1}^{m} u_i^* \frac{\partial g_i}{\partial x_k} = 0; \quad k = 1 \text{ to } n \tag{2.33}$$

$$\frac{\partial L}{\partial v_j} = 0 \quad \text{or} \quad h_j(\mathbf{x}^*) = 0; \quad j = 1 \text{ to } p \tag{2.34}$$

$$\frac{\partial L}{\partial u_i} = 0 \quad \text{or} \quad g_i(\mathbf{x}^*) - s_i^2 = 0; \quad i = 1 \text{ to } m \tag{2.35}$$

3. Compute the switching conditions

$$\frac{\partial L}{\partial s_i} = 0 \quad \text{or} \quad 2u_i^* s_i = 0; \quad i = 1 \text{ to } m \tag{2.36}$$

4. Verify viability of inequality

$$s_i^2 \geq 0 \text{ equivalent to } g_i(\mathbf{x}) \leq 0; \quad i = 1 \text{ to } m \tag{2.37}$$

5. Verify if the Lagrange multipliers of the inequations are nonnegative

$$u_i \geq 0; \quad i = 1 \text{ to } m \tag{2.38}$$

2.5.5 Convexity

A **convex set** S is a collection of $\mathbf{x}$ points, where for all $\mathbf{x}_1$ and $\mathbf{x}_2$, two included points, every segment $\mathbf{x}_1 - \mathbf{x}_2$ is also included in S. A straight segment is always a convex set because all points within the segment obey the definition above. If in a single-variable function the line joining two points is always above the function curve, this is a convex function.

A function of several variables $f(\mathbf{x})$, defined in a convex set, is a convex function if and only if its Hessian is positive half-defined or defined at all points in the set. In the latter case, it is strictly convex.

One can prove a theorem in which a convex programming problem is defined as follows:

It is a general optimization problem whose viable set is given by the following expression:

$$S = \left\{ \mathbf{x} | h_j(\mathbf{x}) = 0, j = 1 \text{ to } p; g_i(\mathbf{x}) = 0, i = 1 \text{ a } m \right\}$$

Thus, a set S is a convex set if g_i are convex and functions h_j are linear. Nonlinear equality constraints always lead to nonconvex sets. Linear equality or inequality constraints always lead to convex sets.

Another theorem states that $f(\mathbf{x}^*)$ is a **local minimum** of a convex function $f(\mathbf{x})$ defined in a viable convex set S, then it is also a **global minimum**.

By virtue of the above theorems, it can be said that if a convex function $f(\mathbf{x})$ is defined in a viable set S, then the necessary KKT conditions are also **sufficient** for a global minimum.

2.6 Examples

2.6.1 Example 1: KKT conditions

Minimize $f(x_1, x_2) = x_1^2 + x_2^2 - 2x_1 - 2x_2 + 2$

    Subject to $g_1 = -2x_1 - x_2 + 4 \leq 0$ $g_2 = -x_1 - 2x_2 + 4 \leq 0$

1. Lagrangian function

$$L = x_1^2 + x_2^2 - 2x_1 - 2x_2 + 2 + u_1\left(-2x_1 - x_2 + 4 + s_1^2\right) + u_2\left(-x_1 - 2x_2 + 4 + s_2^2\right)$$

2. KKT conditions

$$\frac{\partial L}{\partial x_1} = 2x_1 - 2 - 2u_1 - u_2 = 0$$

$$\frac{\partial L}{\partial x_2} = 2x_2 - 2 - u_1 - 2u_2 = 0$$

The partial derivatives of the Lagrangian with respect to the Lagrange multipliers render the original constraint equations:

$$g_1 = -2x_1 - x_2 + 4 + s_1^2 = 0; \quad s_1^2 \geq 0, \quad u_1 \geq 0$$

$$g_1 = -x_1 - 2x_2 + 4 + s_2^2 = 0; \quad s_2^2 \geq 0, \quad u_2 \geq 0$$

The partial derivatives of the Lagrangian with respect to the slack variables always render the same switching conditions:

$$u_i s_i = 0; \quad i = 1, 2$$

Analysis of possible solutions

Case 1: $u_1 = 0$, $u_1 = 0$ leads to $s_1^2 = -1$, $s_2^2 = -1$, impossible.

Case 2: $u_1 = 0$, $s_2 = 0$ leads to $x_1 = 1, 2$; $x_2 = 1, 2$; $u_1 = 0$; $u_2 = 0, 4$; but $s_2^2 = -0, 2$, impossible.

Case 3: $s_1 = 0$, $u_2 = 0$ leads to $s_2^2 = -0, 2$, impossible.

Case 4: $s_1 = 0$, $s_2 = 0$ leads to $x_1 = 4/3$; $x_2 = 4/3$; $u_1 = 2/9$; $u_2 = 2/9$. This is a valid candidate to be a minimum! The objective function value at this point is $f = 2/9$.

2.6.2 Example 2

Two electric generators are interconnected to power a load of at least 60 units of a certain consumer. The operating cost of each generator is a function of its energy output and is given by the expressions below, based on cost per unit.

Formulate the minimum cost problem to determine P_1 and P_2 powers that each generator must provide. Formulate the KKT conditions:

Cost per power unity for generator 1 $- C_1 = 1 - P_1 + P_1{}^2$

Cost per power unity for generator 2 $- C_2 = 1 + 0.6P_2 + P_2{}^2$

Design variables: $x_1 = P_1 \mathrm{e} x_2 = P_2$

Objective function: $f(\mathbf{x}) = C_1 + C_2 = 2 - x_1 + x_1^2 + 0.6x_2 + x_2^2$

Subject to:

$$g_1 = -x_1 - x_2 + 60 \leq 0$$

$$g_2 = -x_1 \leq 0$$

$$g_3 = -x_2 \leq 0$$

1. Lagrangian

$$L = 2 - x_1 + x_1^2 + 0.6x_2 + x_2^2 + u_1\left(-x_1 - x_2 + 60 + s_1^2\right) + u_2\left(-x_1 + s_2^2\right) + u_3\left(-x_2 + s_3^2\right)$$

2. KKT conditions

$$\frac{\partial L}{\partial x_1} = -1 + 2x_1 - u_1 - u_2 = 0$$

$$\frac{\partial L}{\partial x_2} = 0.6 + 2x_2 - u_1 - u_3 = 0$$

The partial derivatives of the Lagrangian with respect to the Lagrange multipliers render the original constraint equations:

$$-x_1 - x_2 + 60 + s_1^2 = 0$$

$$-x_1 + s_2^2 = 0$$

$$-x_2 + s_3^2 = 0$$

3. Switching conditions
The partial derivatives of the Lagrangian with respect to the slack variables always render the same switching conditions:

$$2u_1 s_1 = 0$$

$$2u_2 s_2 = 0$$

$$2u_3 s_3 = 0$$

4. Viability

$$s_i^2 \geq 0; \quad i = 1 \text{ to } 3$$

5. Nonnegativity of Lagrange multipliers

$$u_i \geq 0; \quad i = 1 \text{ to } 3$$

2.6.3 Example 3

The two-bar truss of Figure 2.1 (right triangle 30:40:50 cm) shall be designed to support weight $W = 1,200$ kN at node A, without the bars exceeding the allowable normal stress of 16 kN/cm². The total volume of material should be minimized.

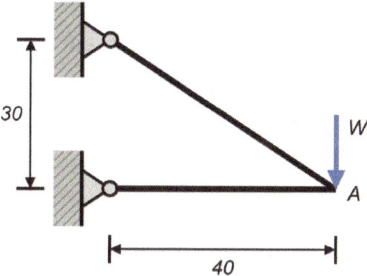

Figure 2.1: Truss Example 2.6.3.

Node A equilibrium:

$$\sum{}^V = 0.6F_1 - 1{,}200 = 0 \quad F_1 = 2{,}000 \text{ kN}$$

$$\sum{}^H = -0.8F_1 - F_2 = 0 \quad F_2 = 1{,}600 \text{ kN}$$

Allowable stress check: $\sigma_i = \dfrac{F_i}{A_i} \leq \bar{\sigma}$

Design variables: $x_1 = A_1$; $x_2 = A_2$, cm^2

Objective function: $f(\mathbf{x}) = 50x_1 + 40x_2$, in cm^2

Subject to $g_1(\mathbf{x}) = \dfrac{2{,}000}{x_1} - 16 \leq 0$; $g_2(\mathbf{x}) = \dfrac{1{,}600}{x_2} - 16 \leq 0$;

$$g_3(\mathbf{x}) = -x_1 \leq 0; \qquad g_4(\mathbf{x}) = -x_2 \leq 0$$

Lagrangian:

$$L = 50x_1 + 40x_2 + u_1\left(\frac{2{,}000}{x_1} - 16 + s_1^2\right) + u_2\left(\frac{1{,}600}{x_1} - 16 + s_2^2\right)$$
$$+ u_3\left(-x_1 + s_3^2\right) + u_4\left(-x_2 + s_4^2\right)$$

KKT conditions:

$$\frac{\partial L}{\partial x_1} = 0 = 50 - u_1\frac{2{,}000}{x_1^2} - u_3$$

$$\frac{\partial L}{\partial x_2} = 0 = 40 - u_2\frac{1{,}600}{x_2^2} - u_4$$

The partial derivatives of the Lagrangian with respect to the Lagrange multipliers render the original constraint equations. The partial derivatives of the Lagrangian with respect to the slack variables always render the same switching conditions:

$$u_i s_i = 0; \quad u_i \geq 0; \quad g_i + s_i^2 = 0; \quad s_i^2 \geq 0; \quad i = 1-4$$

Cases $s_3 = s_4 = 0$ lead to null transverse sections, that is, physically unacceptable.

Other cases:

Case 1: $u_1 = 0$; $u_2 = 0$; $u_3 = 0$; $u_4 = 0$; result in $50 = 0$ and $40 = 0$, that is, absurd!

Case 2: $s_1 = 0$; $u_2 = 0$; $u_3 = 0$; $u_4 = 0$; result in $50 = 0$, that is, absurd!

Case 3: $u_1 = 0$; $s_2 = 0$; $u_3 = 0$; $u_4 = 0$; result in $40 = 0$, that is, absurd!

Case 4: $s_1 = 0$; $s_2 = 0$; $u_3 = 0$; $u_4 = 0$; result in $x_1^* = 125$ cm^2, $x_2^* = 100$ cm^2,
$u_1 = 0.391$ e $u_2 = 0.25$, a candidate to local minimum.

As $f(\mathbf{x})$, $g_3(\mathbf{x})$, and $g_4(\mathbf{x})$ are linear and the Hessian matrices of $g_1(\mathbf{x})$ and $g_2(\mathbf{x})$ are positive semidefinite, the problem is convex and this solution fulfills the sufficient conditions to be a global minimum.

2.6.4 Example 4

The cantilever beam of Figure 2.2, rectangular cross-section b (the width) x d (the height must not exceed 2 times the width), has a span of $L = 400/15$ cm and supports vertical force $V = 150$ kN at its free end, without exceeding the allowable normal stress 1 kN/cm^2, and the allowable shear stress is 0.2 kN/cm^2. The total volume of a material should be minimized.

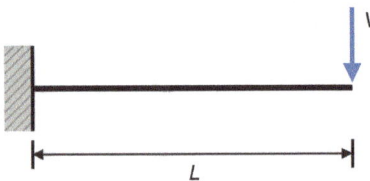

V

L

Figure 2.2: Beam Example 2.6.4.

Maximum bending moment: $150 \times 400/15 = 4{,}000$ kN cm

Maximum shear force: 150 kN

Normal stress check: $\sigma = \dfrac{6M}{bd^2} \leq \bar{\sigma}$

Shear stress check: $\tau = \dfrac{3}{2}\dfrac{V}{bd} \leq \bar{\tau}$

Design variables: $x_1 = b$; $x_2 = d$, cm

Objective function: $f(\mathbf{x}) = x_1 x_2$, cm^2

Subject to

$$g_1(\mathbf{x}) = \frac{6 \times 4{,}000}{x_1 x_2^2} - 1 \leq 0; \qquad g_2(\mathbf{x}) = \frac{3 \times 150}{2 x_1 x_2} - 0.2 \leq 0;$$

$$g_3(\mathbf{x}) = x_2 - 2x_1 \leq 0; \qquad g_4(\mathbf{x}) = -x_1 \leq 0; \qquad g_5(\mathbf{x}) = -x_2 \leq 0$$

Lagrangian:

$$L = x_1 x_2 + u_1 \left(\frac{2{,}400}{x_1 x_2^2} - 1 + s_1^2 \right) + u_2 \left(\frac{225}{x_1 x_2} - 0{,}2 + s_2^2 \right) + u_3 \left(x_2 - 2x_1 + s_3^2 \right) + u_4 \left(-x_1 + s_4^2 \right)$$
$$+ u_5 \left(-x_2 + s_5^2 \right)$$

KKT conditions:

$$\frac{\partial L}{\partial x_1} = 0 = x_2 - u_1 \frac{2{,}400}{x_1^2 x_2^2} - u_2 \frac{225}{x_1^2 x_2} - 2u_3 - u_4$$

$$\frac{\partial L}{\partial x_2} = 0 = x_1 - u_1 \frac{4{,}800}{x_1^2 x_2^3} - u_2 \frac{225}{x_1 x_2^2} + u_3 - u_5$$

The partial derivatives of the Lagrangian with respect to the Lagrange multipliers render the original constraint equations. The partial derivatives of the Lagrangian with respect to the slack variables always render the same switching conditions:

$$u_i s_i = 0; \quad u_i \geq 0; \quad g_i + s_i^2 = 0; \quad s_i^2 \geq 0; \quad i = 1 - 5$$

Cases where $s_4 = 0$ or $s_5 = 0$ or $s_5 = s_4 = 0$ are not possible. Thus, we must have $u_5 = u_4 = 0$ for all cases.

Case 1: $u_1 = 0$; $u_2 = 0$; $u_3 = 0$; $u_4 = 0$; $u_5 = 0$; resulting in $b = 0$ and $d = 0$, that is not acceptable.

Case 2: $u_1 = 0$; $u_2 = 0$; $s_3 = 0$; $u_4 = 0$; $u_5 = 0$; resulting in $b = 0$ and $d = 0$, that is not acceptable.

Case 3: $u_1 = 0$; $s_2 = 0$; $u_3 = 0$; $u_4 = 0$; $u_5 = 0$; result in infinite valid solutions in the ranges of values:

$$23.717 \leq b \leq 52.734 \text{ cm}; \quad 21.333 \leq d \leq 47.433 \text{ cm, sujeito a } bd = 1125 \text{ cm}$$

Case 4: $s_1 = 0$; $u_2 = 0$; $u_3 = 0$; $u_4 = 0$; $u_5 = 0$ have no consistent solutions.

Case 5: $u_1 = 0$; $s_2 = 0$; $s_3 = 0$; $u_4 = 0$; $u_5 = 0$ is a limit of case 3, namely $b = 23.717$ cm and $d = 47.434$ cm.

Case 6: $s_1 = 0$; $s_2 = 0$; $u_3 = 0$; $u_4 = 0$; $u_5 = 0$ is a limit of case 3, namely $b = 52.374$ cmand $d = 21.333$ cm.

Case 7: $s_1 = 0$; $u_2 = 0$; $s_3 = 0$; $u_4 = 0$; $u_5 = 0$ results in $u_3 < 0$, which is not valid.

Case 8: $s_1 = 0$; $s_2 = 0$; $s_3 = 0$; $u_4 = 0$; $u_5 = 0$ results in a system of three equations with two unknowns that has no solutions.

2.6.5 Example 5: proposed

Determine the minimum mass of an aluminum tripod of height H to withstand a vertical load $V = 60$ kN. The base is an equilateral triangle with sides $B = 120$ cm. The bars have solid circular cross section of diameter D and must not exceed the allowable stress of the material nor the critical Euler buckling load with safety coefficient 2. The ranges of values of the design variables are allowable compressive stress 150 MPa, modulus of elasticity 75 GPa, and density 2,800 kg/m^3.

Answers: $H = 50$ cm; $D = 3.42$ cm; minimum mass = 6.6 kg.

2.7 Functionals and their maxima and minima

In this presentation, we will try to maintain a certain parallelism between variational calculus and the classical differential calculus.

Let B be a vector space of functions. A functional is an application Π that associates with each element f of B, a single element y of **R**. It is denoted as $\Pi{:}B \mapsto R$, so that if $f \in B$ then $y = \Pi(f)$.

A functional $\Pi{:}B \mapsto R$ is said **convex** if

$$\Pi((1-\theta)f_a + \theta f_b) \leq (1-\theta)\Pi(f_a) + \theta\Pi(f_b) \qquad f_a, f_b \in \text{B}, \qquad \forall\theta \in [0,1]. \qquad (2.39)$$

A functional $\Pi{:}B \mapsto R$ is said **strictly convex** if

$$\Pi((1-\theta)f_a + \theta f_b) < (1-\theta)\Pi(f_a) + \theta\Pi(f_b) \qquad \forall f_a, f_b \in \text{B}, \qquad \forall\theta \in (0,1) \qquad (2.40)$$

Let us define $V_h(f_0) = \{\Omega \subset \text{B} | \forall f \in \Omega, d(f, f_0) < \varepsilon\ \}$, a neighborhood of f_0.

A functional $\Pi{:}B \mapsto R$ reaches a **local minimum** in f_0 if there is a neighborhood of f_0 where

$$\Pi\ (f) \geq \Pi(f_0) \qquad \forall f \in V_h(f_0). \qquad (2.41)$$

This minimum is said **global** if

$$\Pi\ (f) \geq \Pi(f_0) \qquad \forall f \in \text{B} \qquad (2.42)$$

This minimum is said **strict** if

$$\Pi\ (f) > \Pi(f_0) \qquad \forall f \in V_h(f_0)\ |\ d(f, f_0) \neq 0 \qquad (2.43)$$

A functional $\Pi{:}B \mapsto R$ reaches a **local maximum** in f_0 if there is a neighborhood of f_0 where

$$\Pi\ (f) \leq \Pi(f_0) \qquad \forall f \in V_h(f_0) \qquad (2.44)$$

This maximum is said **global** if

$$\Pi\,(f) \le \Pi(f_0) \qquad \forall f \in \mathrm{B} \tag{2.45}$$

This maximum is said **strict** if

$$\Pi\,(f) < \Pi(f_0) \qquad \forall f \in V_h(f_0) \mid d(f,f_0) \ne 0 \tag{2.46}$$

Note: Convex functions have at least a global minimum. When they are strictly convex, this minimum not only exists but is unique. It is said that $\Pi(f_0)$ is an **extreme** of Π and that f_0 is a **stationary function**.

Chapter 3
Graphic method

A class of problems reasonably common in practice is that of only two design variables, or, as can also be said, two degrees of freedom.

In this case, it is possible to represent all possible values of these two variables, including constraints, and plot isovalues of the objective function in a plane and visually determine the solution of the corresponding optimization problem, without using mathematical tools other than the simplest analytical geometry.

3.1 Examples

3.1.1 Example E1 of Chapter 1

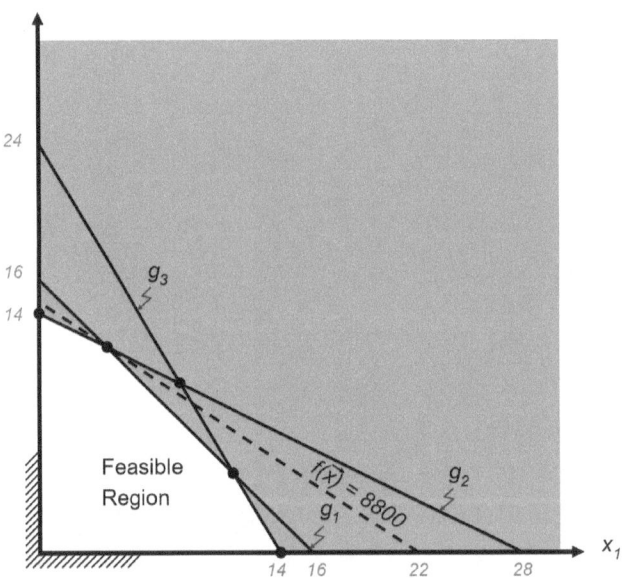

Figure 3.1: Graphic solution of Example 3.1.

Design variables:

x_1 = number of A toys, x_2 = number of B toys

Objective function (profit), to be maximized: $F(\boldsymbol{x}) = 400x_1 + 600x_2$

https://doi.org/10.1515/9783110625622-003

Inequality constraints:

$$x_1 + x_2 \leq 16 \quad \Rightarrow \quad g_1(\boldsymbol{x}) = x_1 + x_2 - 16 \leq 0 \qquad \text{(shipping)}$$

$$x_1/28 + x_2/14 \leq 1 \quad \Rightarrow \quad g_2(\boldsymbol{x}) = x_1/28 + x_2/14 - 1 \leq 0 \quad \text{(production)}$$

$$x_1/14 + x_2/24 \leq 1 \quad \Rightarrow \quad g_3(\boldsymbol{x}) = x_1/14 + x_2/24 - 1 \leq 0 \quad \text{(sales)}$$

The plot of these functions is shown in Figure 3.1. The isoprofit line that gives the maximum value and meets all constraints is $400x_1 + 600x_2 = 8{,}800$. The available plotting software, such as Microsoft Excel, greatly facilitate this form of resolution.

3.1.2 Column supporting compressive axial loading

A column, shown in Figure 3.2, is $L = 5$ m tall, which is clamped at its base and free at the upper end. The section is tubular of average radius R and wall thickness t.

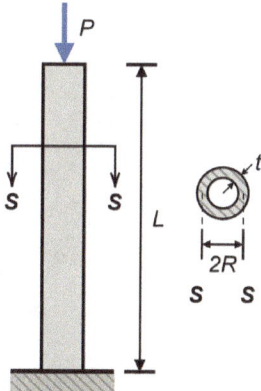

Figure 3.2: Column of Example 3.2.

It supports an axial compressive force, $P = 10$ MN. Determine its minimum mass, knowing that it is made up of steel ($E = 207$ GPa, density $\rho = 7{,}833$ kg/m^3, allowable strength $\sigma_a = 248$ MPa).

Design variables: R and t

Objective function: $f(R, t) = 2\rho L\pi Rt$, mass in kg

Inequality constraints:

$$g_1(R, t) = \frac{P}{2\pi Rt} - \sigma_a \leq 0 \qquad \text{(allowable strength)}$$

$$g_2(R, t) = P - \frac{\pi^3 ER^3 t}{4L^2} \leq 0 \qquad \text{(buckling load)}$$

$$g_3(R, t) = -R \leq 0, \qquad g_4(R, t) = -t \leq 0$$

It is a nonlinear minimization problem with constraints. Since there are only two design variables, it is possible to solve it by the graphical method in the Cartesian plane $R \times t$. There are an infinite number of solutions. All points in the curve segment $g_1(R, t) = \dfrac{P}{2\pi R t} - \sigma_a = 0$ above its intersection with curve $g_2(R, t) = P - \dfrac{\pi^3 E R^3 t}{4L^2} = 0$ are solutions rendering column mass 1,579 kg. Particularly, this intersection point $R = 0.1575$ m, and $t = 0.0405$ m is adopted as optimum, among all infinite possible solutions.

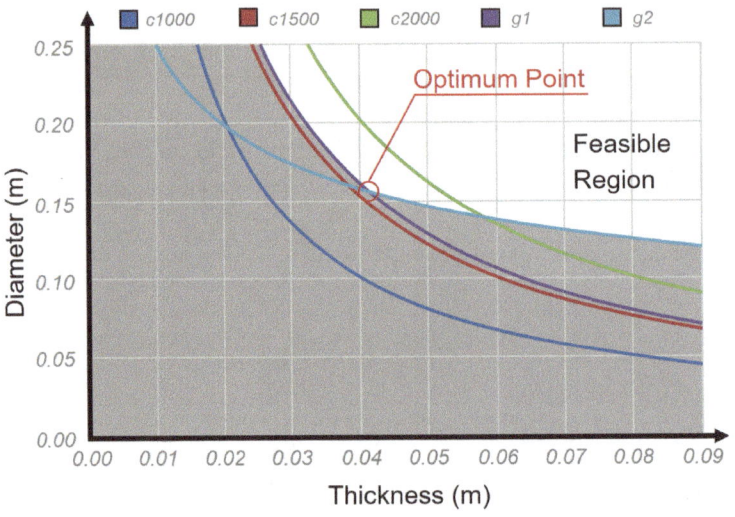

Figure 3.3: Graphic solution of Example 3.2.

See the graphical solution in Figure 3.3. In this figure, we have isomass curves c1000, c1500, and c2000 for, respectively, the costs of 1,000, 1,500, and 2,500 kg, besides restrictions g_1 and g_2. The objective function is also called the cost function because of its usual application to cost minimization problems.

3.1.3 Flexed beam

Consider the simply supported beam having length $L = 10$ m, as shown in Figure 3.4. It is subject to a concentrated load $P = 20$ tf applied at mid-span. The cross section is rectangular with dimensions $b_w \times h$. The material composing the beam has allowable stress $\sigma_a = 20$ MPa and density $\sigma = 2,500$ kg/m^3. The optimization (mass minimization) problem is defined as follows.

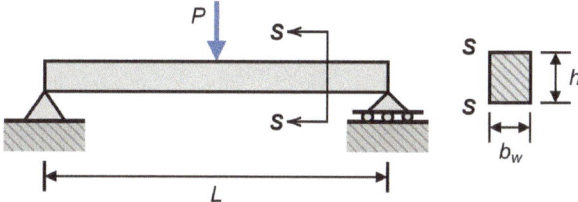

Figure 3.4: Simply supported beam, subject to concentrated load at mid-span.

Design variables: b_w and h.

Objective function: $f(b_w, h) = \rho L b_w h$, mass in kg

Inequality constraints:

$$g_1(b_w, h) = \frac{3}{2} \frac{PL}{b_w h^2} - \sigma_a \le 0 \quad \text{(allowable strength)}$$

$$g_2(b_w, h) = b_w - h \le 0 \qquad \text{(height larger than width)}$$

$$g_3(b_w, h) = -b_w + 0.2 \le 0 \quad \text{(minimum width)}$$

$$g_4(b_w, h) = -h + 0.2 \le 0 \qquad \text{(minimum height)}$$

By plotting the graphs of the objective function and the constraints, we obtain Figure 3.5. It is observed that the optimal value lies at the intersection of the constraints g_1 and g_3. In this case, it is said that g_1 and g_3 are active constraints. The other constraints g_2 and g_4 are far from the viable region and do not interfere with the solution of the problem. Some search algorithms disregard constraints that are not active or ε-active in the calculation of the gradient of the constraints for the determination of the optimal solution. Another observation is that g_4 is redundant as g_2 and g_3 also define a minimum value for h equals 0.2. The definition of redundant constraints increases the computational time involved and can negatively interfere in the search process, so they should be avoided.

The optimum value of total mass is 4,333 kg at $b_w = 0.2$ m and $h = 0.867$ m.

3.1.4 Example

Two electric generators are interconnected to provide a load of at least 60 units of power to a certain consumer. The operating cost of each generator is a function of its energy output given by the below expressions, based on cost per power unit. Formulate the minimum cost problem to determine P_1 and P_2 powers that each generator must provide. Determine the solution graphically.

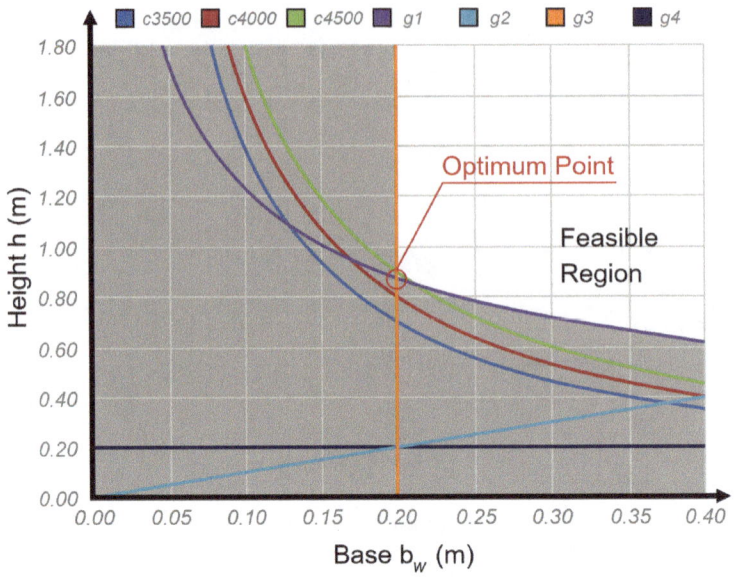

Figure 3.5: Graphic solution of Example 3.3.

Cost of power unit for generator: $1 - C_1 = 1 - P_1 + P_1{}^2$
Cost of power unit for generator: $2 - C_2 = 1 + 0.6P_2 + P_2{}^2$
Design variables: $x_1 = P_1$ and $x_2 = P_2$
Objective function: $f(\mathbf{x}) = C_1 + C_2 = 2 - x_1 + x_1^2 + 0.6x_2 + x_2^2$
Subject to:

$$g_1 = -x_1 - x_2 + 60 \leq 0$$

$$g_2 = -x_1 \leq 0$$

$$g_3 = -x_2 \leq 0$$

In Figure 3.6, the solution of the problem is shown via the function graphs. The optimum point is $\mathbf{x}^{\star T} = [30.4\ 29.6]$ where the value of the objective function is $f(\mathbf{x}^\star) = 1{,}790$.

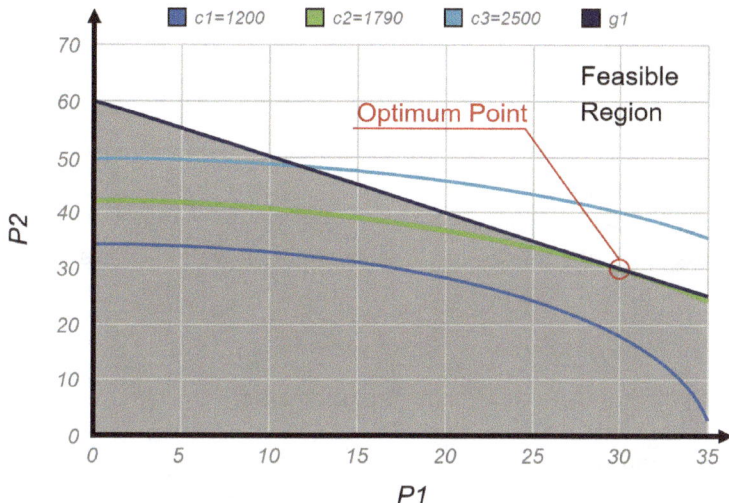

Figure 3.6: Graphic solution of Example 3.4.

Chapter 4
Linear programming

An optimization problem involving only linear functions of the design variables is also called a linear programming problem.

Linear programming is usually considered as an operational research method, but there is a very large series of applications. The problem that will be exposed can be expressed in its standard form as follows:

$$\text{Minimize } f(\mathbf{x}) = \mathbf{c}^T \mathbf{x} \quad \text{(objective function)}$$

$$\text{Subject to } \mathbf{Ax} = \mathbf{b} \quad \text{(constraint equations), where } \mathbf{x} \geq 0$$

$\mathbf{x}$ is the column vector of the design variables to be determined. The given constants of the system, also known as **available resources**, are provided by the vector column $\mathbf{b}$, a matrix $\mathbf{A}$, and a column vector $\mathbf{c}$. All the constraint equations and the objective function to be minimized are in linear form.

Again, the problem represents the need to minimize a linear function, the objective function, subject to satisfying a linear system of equalities. Although it has been given the "standard" form, many other forms of this problem may appear, which are converted to it. For example, constraints may initially be inequalities and they can be converted into equalities by adding or subtracting additional variables called **slack variables**. The goal may be to maximize the function rather than minimize it. Again this is obtained by changing the signals of the coefficients.

As said, all n inequality constraint equations must be transformed into equality equations by including unknown positive slack variables. Thus, the size of the design variable vector $\mathbf{x}$ will be the number m of the real design variables we wish to determine in order to minimize f, plus the number n of slack variables. In consequence, matrix $\mathbf{A}$ is $n \times m$, the "available resources" vector $\mathbf{b}$ will be $n \times 1$, and vector $\mathbf{c}$ will be $(n + m) \times 1$, the same size of vector $\mathbf{x}$.

Some practical examples where linear programming can be applied are:
- problem of food diets in hospitals, requiring reduction of food costs, while remaining offering the best diet;
- problem of reduction of standard loss in industries;
- problem of optimizing profit, subject to material availability restrictions;
- problem of optimization of telephone call routines;
- minimization of transportation costs.

https://doi.org/10.1515/9783110625622-004

4.1 The SIMPLEX method

A powerful numerical method for solving linear programming problems is called SIM-PLEX, one of the first to become available and popular when digital electronic stored program computers were introduced in the second half of the twentieth century.

For the exposition of the algorithm, we will use the same example of profit maximization solved graphically in Section 3.1. In the first phase, the equations of inequality constraints are transformed into equations of equality constraints by the introduction of additional variables that represent the surplus of resources existing in each of them, called **slack variables**. The problem is

$$\text{Minimize } f(x) = -400x_1 - 600x_2, \text{ subject to}$$

$$x_1 + x_2 + x_3 = 16$$

$$x_1/28 + x_2/14 + x_4 = 1$$

$$x_1/14 + x_2/24 + x_5 = 1$$

In array form, we have the design variables vector $x = \begin{bmatrix} x_1 & x_2 & \cdots & x_5 \end{bmatrix}^T$, the objective function $f(\mathbf{x}) = \mathbf{c}^T\mathbf{x}$, where $\mathbf{c} = \begin{bmatrix} c_1 & c_2 & \cdots & c_5 \end{bmatrix}^T = \begin{bmatrix} -400 & -600 & \cdots & 0 \end{bmatrix}^T$ and the constraint equations $\mathbf{Ax} = \mathbf{b}$. The components a_{ij} of matrix $\mathbf{A}$, $m \times n$ (in the case $m = 3$ and $n = 5$), are the coefficients of the constraint equations and $\mathbf{b} = \begin{bmatrix} b_1 & b_2 & b_3 \end{bmatrix}^T = \begin{bmatrix} 16 & 1 & 1 \end{bmatrix}^T$.

Since the matrix $\mathbf{A}$ is 3×5, that is, $m < n$, there is no single solution. We introduce the concept of basic solution, in which $n - m$ variables are annulled (in this case, two), called nonbasic variables, and the others are called basic variables, allowing the solution of the remaining system (in this case, 3×3). Each of these basic solutions corresponds to a vertex of the polygon of Figure 4.1. As one can see, of the 10 basic solutions possible in this case, some are viable (with respect to all restrictions) and others are not feasible. The inspection of all these possible basic solutions is a brute-force procedure to solve the problem. In a case of large size problem, it becomes economically unfeasible.

The SIMPLEX method is organized into tables called Tableau, each representing one basic solution. The passage from one solution to another is done in an intelligent way, and there is a criterion to know when the solution of the optimization problem is reached. The starting Tableau is shown in Table 4.1.

In this solution, the basic variables $x_3 = 16$, $x_4 = 1$, $x_5 = 1$, and $x_1 = 0$, $x_2 = 0$ are nonbasic, obviously leading to $f = 0$ (see Figure 4.1). By the method, to examine a new basic solution, one of the basic variables must become nonbasic and a nonbasic variable must become basic. The criterion for this is to adopt the column that corresponds to the lowest cost (the second column, cost -600) and the line corresponding to the lowest positive ratio b_i/a_{i2}. Element $a_{22} = 1/14$ is the new pivot of the procedure. In the numerical method, it is customary to make this a unitary pivot

Table 4.1: The first Tableau.

Basic variable	x_1	x_2	x_3	x_4	x_5	b	Ratio b_i/a_{i2}
x_3	1	1	1	0	0	16	16
x_4	1/28	1/14	0	1	0	1	14
x_5	1/14	1/24	0	0	1	1	24
Cost	−400	−600	0	0	0	$f - 0$	

dividing this line by itself. In addition, this line, multiplied by an appropriate number, is subtracted from the other lines to zero the coefficients in column 2. The result is the second Tableau (see Table 4.2 and Figure 4.1).

Table 4.2: The second Tableau.

Basic variable	x_1	x_2	x_3	x_4	x_5	b	Ratio b_i/a_{i1}
x_3	1/2	0	1	−14	0	2	4
x_2	1/2	1	0	14	0	14	28
x_5	17/336	0	0	−7/12	1	7	140/17
Cost	−100	0	0	8,400	0	$f + 8,400$	

We select now the column that corresponds to the lowest cost (the first column, cost − 100) and the line corresponding to the smallest positive ratio b_i/a_{i1}. Element $a_{11} = 1/2$ is the new pivot of the procedure. This pivot is made unitary and we divide this line by itself and subtract it, multiplied by an appropriate number, from the other lines to zero the coefficients of column 1. The result is the third Tableau (Table 4.3).

One can prove that the process reaches the minimum when the reduced values from the cost line for the nonbasic variables are nonnegative (instead of decreasing, they increase the cost). The solution to the problem are, therefore, basic variables $x_1 = 4$, $x_2 = 12$, $x_5 = 3/14$, and $x_3 = 0$, $x_4 = 0$ the nonbasic variables, resulting $f = -8,800$.

In Figure 4.1, the points corresponding to each of the tables are marked. Notice that the algorithm moves the solution along the vertices of the polygon that defines the feasible region of the problem (viable domain) until the optimal value is obtained.

There are particular cases, but their examination goes beyond the scope of this book and should be sought in the operational research literature.

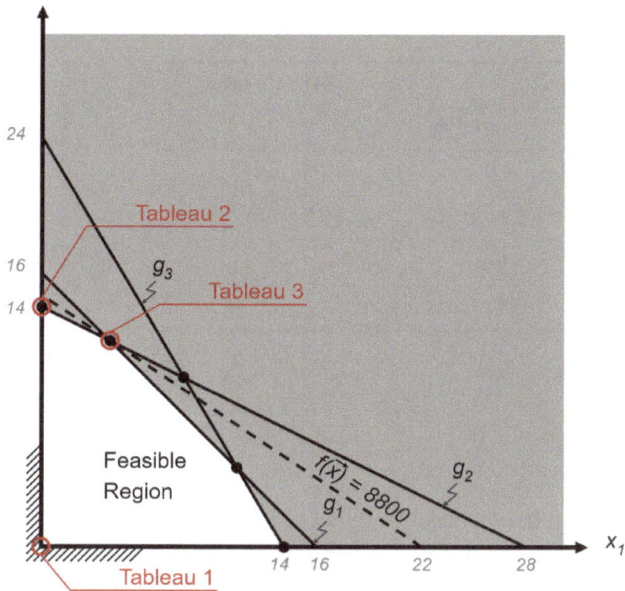

Figure 4.1: Graphic representation of Example 4.1.

Table 4.3: The third Tableau.

Basic variable	x_1	x_2	x_3	x_4	x_5	b	Ratio b_i/a_{i1}
x_1	1	0	2	−28	0	4	Not necessary
x_2	0	1	−1	28	0	12	Not necessary
x_5	0	0	−17/168	5/6	1	3/14	Not necessary
Cost	0	0	200	5,600	0	$f + 8,800$	

4.2 Example

A refinery receives a fixed amount of gross natural gas in m^3 per week. It is processed into two types: common and special gas qualities, each consuming a certain time and giving a certain profit per ton processed. Only one of the product types can be processed at a time. The refinery works 80 h per week and its storage capacity is restricted for each type of product. How much gas of each type should be processed for maximum profit? The data available to the manager to solve the linear programming problem are summarized in Table 4.4.

Table 4.4: Problem data.

Resources	Product		Available resources
	Common gas	**Special gas**	
Gross gas	7 m³/ton	11 m³/ton	77 m³/week
Production time	10 h/ton	8 h/ton	80 h/week
Storage	9 ton	6 ton	
Profit	$ 150/ton	$ 175/ton	

Design variables:

x_1: amount of common gas to be produced

x_2: amount of special gas to be produced

Objective function (profit): $F(\mathbf{x}) = 150x_1 + 175x_2$

Constraints with *slack variables*:

$$g_1 = 7x_1 + 11x_2 + x_3 = 77$$

$$g_2 = 10x_1 + 8x_2 + x_4 = 80$$

$$g_3 = x_1 + x_5 = 9$$

$$g_4 = x_2 + x_6 = 6$$

Applying SIMPLEX, the first Tableau is shown in Table 4.5.

Table 4.5: The first Tableau.

Basic variable	x_1	x_2	x_3	x_4	x_5	x_6	b	Ratio b_i/a_{i2}
x_3	7	11	1	0	0	0	77	7
x_4	10	8	0	1	0	0	80	10
x_5	1	0	0	0	1	0	9	∞
x_6	0	1	0	0	0	1	6	6
Cost	−150	−175	0	0	0	0	$f - 0$	

The pivot is $a_{42} = 1$ as the second column is of lower cost (−175), and the fourth line has the smallest positive ratio b_i/a_{i2}.

This pivot is already unitary, as required by the method. By zeroing the coefficients of this column above and below this line, one arrives at the second Tableau (Table 4.6).

Table 4.6: The second Tableau.

Basic variable	x_1	x_2	x_3	x_4	x_5	x_6	b	Ratio b_i/a_{i1}
x_3	7	0	1	0	0	-11	11	1.57143
x_4	10	0	0	1	0	-8	32	3.2
x_5	1	0	0	0	1	0	9	9
x_2	0	1	0	0	0	1	6	∞
Cost	-150	0	0	0	0	175	$f+1,050$	

The new pivot is $a_{11} = 7$ as the first column gives the lower cost (−150), and the first line the smallest positive ratio b_i/a_{i2}.

Making this unitary pivot and zeroing the coefficients below that line, one arrives at the third Tableau (Table 4.7).

Table 4.7: The third Tableau.

Basic variable	x_1	x_2	x_3	x_4	x_5	x_6	b	Ratio b_i/a_{i2}
x_1	1	0	0.1429	0	0	-1.571	1.57143	
x_4	0	0	-0.1429	1	0	7.7143	16.2857	2.1111
x_5	0	0	-0.1429	0	1	1.571	7.4286	4.7286
x_2	0	1	0	0	0	1	6	6
Cost	0	0	21.4256	0	0	-60.71	$f+1,286$	

The new pivot is $a_{26} = 7.7143$ as the sixth column corresponds to the lower cost (−60,71), and the second line to the smallest positive ratio b_i/a_{i2}.

Making this unitary pivot and zeroing the coefficients below that line, one arrives at the fourth Tableau (Table 4.8).

This is the last Tableau because in the line of costs there are only nonnegative values. The solution is 4.8889 tons of common gas and 3.8888 tons of special gas per week, resulting in a maximum profit of $ 1,413.80. Another interesting result for the manager is: there is a storage capacity surplus of 2.1111 tons of the special gas and 4.11111 tons of the common gas.

In Figure 4.2, the points corresponding to each Tableau are marked. In addition to restrictions $g1$ to $g4$, the graphs of the objective function are shown for the values $c1 = 1,000$ (objective function value equals 1,000), $c2 = 1,413.80$ (objective function value equals 1,413.80), and $c3 = 2,000$ (objective function value equals 2,000). Notice that in this case also the simplex algorithm moves the solution along the vertices of

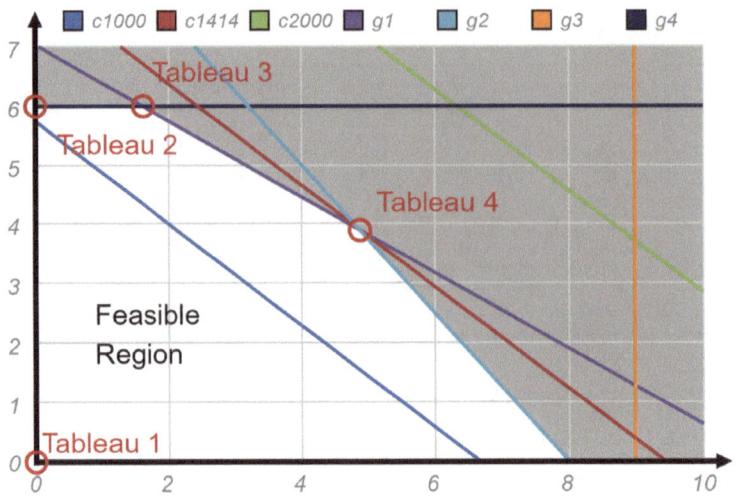

Figure 4.2: Graphic solution of Example 4.2.

Table 4.8: The fourth and last Tableau.

Basic variable	x_1	x_2	x_3	x_4	x_5	x_6	b	Ratio b_i/a_{i2}
x_1	1	0	0.1481	0.2037	0	0	4.8889	
x_6	0	0	−0.1852	0.1296	0	1	2.1111	
x_5	0	0	0.1481	−0.204	1	0	4.1111	
x_2	0	1	0.1852	−0.13	0	0	3.8889	
Cost	0	0	10.1852	7.8704	0	0	$f + 1,413.8$	

the polygon that defines the feasible region of the problem (viable domain) until the optimal value is obtained.

Chapter 5
Using MATLAB optimization toolbox

In this chapter, the use of the MATLAB optimization toolbox is presented. This program is an extremely user-friendly numerical analysis and simulation tool. It facilitates, in particular, the use of vectors and matrices, hence its name, which can be translated as Matrix Lab.

The MATLAB program installation package can include several toolboxes each dedicated to a specific application, bringing together preprogrammed algorithms from that area. Here, the optimization toolbox will be used.

The basic ideas of the program and its usual commands are NOT given in this book. To learn that, the reader should look for one of the many good books on the subject and we will consider he knows how to use the basic tools of the program.

5.1 Optimization functions of MATLAB toolbox

The MATLAB optimization toolbox includes the following six basic functions.

fminbnd – optimization of a function of a single variable within a fixed range, that is:

$$\text{find } x \in [x_L \ x_U] \text{ that minimizes } f(x)$$

fminunc, fminsearch – minimization **without restriction** of several variable functions, that is:

$$\text{find } \mathbf{x} \text{ that minimizes } f(\mathbf{x})$$

fmincon – minimization of several variable functions **with constraints**, that is:

$$\text{find } \mathbf{x} \text{ that minimizes } f(\mathbf{x}) \text{ subject to}$$

$$\mathbf{Ax} \leq \mathbf{b}, \quad \mathbf{Nx} = \mathbf{e}$$

$$g_i(\mathbf{x}) \leq 0, \quad i = 1, \ldots, m$$

$$h_j(\mathbf{x}) \leq 0, \quad j = 1, \ldots, p$$

$$x_{iL} \leq x_i \leq x_{iU}$$

linprog – linear programming, that is:

$$\text{find } \mathbf{x} \text{ that minimizes } f(\mathbf{x}) = \mathbf{c}^T\mathbf{x} \text{ subject to}$$

$$\mathbf{Ax} \leq \mathbf{b}, \quad \mathbf{Nx} = \mathbf{e}$$

https://doi.org/10.1515/9783110625622-005

quadprog – quadratic programming, that is:

$$\text{find } \mathbf{x} \text{ that minimizes } f(\mathbf{x}) = \mathbf{c}^T\mathbf{x} + \frac{1}{2}\mathbf{x}^T\mathbf{H}\mathbf{x} \text{ subject to}$$

$$\mathbf{A}\mathbf{x} \le \mathbf{b}, \ \mathbf{N}\mathbf{x} = \mathbf{e}$$

These functions issue outputs that the user must verify:

$\mathbf{x}$ is the vector or solution matrix found by the optimization function used. If ExitFlag > 0, then $\mathbf{x}$ is the solution; if not, $\mathbf{x}$ is the last value obtained by the function.

FunValue – the objective function value, ObjFun, for solution $\mathbf{x}$.

ExitFlag – the output condition code of the optimization function. If ExitFlag > 0, then $\mathbf{x}$ is the solution; if not, $\mathbf{x}$ is the last value obtained by the function; if null, the maximum number of evaluations of the objective function was reached; if negative, the routine did not converge to a solution.

Output – it is an output structure of the optimization function that contains information about the process.

(Output.iterations) – this provides the number of evaluations of the objective function.

(Output.algorithm) – the name of the algorithm is used to solve the problem.

It is strongly suggested that the problem should be formulated as a *main program* that calls a MATLAB optimization subroutine which in turn calls a subroutine that contains the objective function and constraint function, if necessary. The general form of the command is

[x, FunValue, ExitFlag, Output] = fminX ('ObjFun', . . ., 'ConFun', options, other parameters)

where ObjFun is the name of a subroutine, an .m file, which contains the objective function to be called by the optimization function. This subroutine may also contain explicitly the objective function gradient, if necessary. If this is not given, MARLAB will numerically compute the gradient, via finite differences. options are parameters of the optimization function the user wants to change in relation to the pre-programmed default values. After that it is possible to pass other problem related parameters from the main program to the subroutines. In problems with constraints, a subroutine ConFun containing these constraints and their gradients is also required.

Next, we will only detail the use of three of these basic functions: fminserch for multivariable unconstrained optimization problems, fmincon for constrained multivariable optimization problems and linprog for linear programming. These are the three most popular features of the MATLAB Optimization Toolbox.

5.1.1 Multivariable unconstrained optimization problems

We strongly recommend to write a main program MATLAB script where we may first furnish a set of options that may differ from the default setting of this function such as, typically,

$$\text{options} = \text{optimset ('LargeScale', 'off', 'TolCon', 1e} - 8, \text{'TolX', 1e} - 8)$$

Next, we provide lower and upper bound for the design variables, such as minimum and maximum thickness of plates etc. This may be line matrices, if we have a multivariable problem, or a scalar value otherwise. For example for a 2 variable problem:

$$\text{Lb} = [\text{MinX1 MinX2}]; \text{Ub} = [\text{MaxX1 MaxX2}]$$

It is also necessary to provide an initial design, a line matrix with initial values of the design variables:

$$\text{x0} = [\text{InitX1 InitX2}]$$

The fundamental feature of this main program script is the optimization function call:

$$[\text{x, FunVal, ExitFlag, Output}] = \text{fminsearch ('ObjFun', x0, options, Prob\_data)}$$

Here, Prob_data may be an optional set of physical values we wish to pass to the optimization function, in the same fashion it would be done in any computer code from a main program to the subroutines. Of course, although this is an elegant procedure, it is not necessary, as we can write those values in the subroutines themselves.

Any information not used, such as the absence of the options matrix, in a particular problem, must be substituted by an empty matrix [ ].

5.1.2 Multivariable constrained optimization problems

We strongly recommend to write a main program MATLAB script where we may first furnish a set of options that may differ from the default setting of this function such as, typically,

$$\text{options} = \text{optimset ('LargeScale', 'off', 'TolCon', 1e} - 8, \text{'TolX', 1e} - 8)$$

Next, we provide lower and upper bound for the design variables, such as minimum and maximum thickness of plates. This may be line matrices, if we have a multivariable problem, or a scalar value otherwise. For example for a 2 variable problem:

$$\text{Lb} = [\text{MinX1 MinX2}]; \text{Ub} = [\text{MaxX1 MaxX2}]$$

It is also necessary to provide an initial design, a line matrix with initial values of the design variables:

$$x0 = [\text{InitX1 InitX2}]$$

The fundamental feature of this main program script is the optimization function call:

$$[x, \text{FunVal}, \text{ExitFlag}, \text{Output}] = \ldots \text{fmincon}('\text{ObjFun}', x0, A, b, \text{Aeq}, \text{beq}, \text{Lb}, \text{Ub},$$

$$'\text{ConFun}', \text{options}, \text{Prob\_data})$$

Here, A is the matrix of coefficients of possible **linear inequality constraint equations,** b the right-hand side constants vector of these equations (the so-called available resources). Aeq is the matrix of coefficients of possible **linear equality constraint equations,** b the right-hand side constants vector of these equations.

Prob_data may be an optional set of physical values we wish to pass to the optimization function, in the same fashion it would be done in any computer code from a main program to the subroutines. Of course, although this is an elegant procedure, it is not necessary, as we can write those values in the subroutines themselves.

Any information not used, such as the absence of linear constraints, in a particular problem, must be substituted by an empty matrix [ ], as:

$$x, \text{FunVal}, \text{ExitFlag}, \text{Output}] = \ldots \text{fmincon}('\text{ObjFun}', x0, [\,], [\,], [\,], [\,], \text{Lb}, \text{Ub},$$

$$'\text{ConFun}', \text{options}, \text{Prob\_data})$$

5.1.3 Linear programming problems

A linear programming problem is set as
  Minimize an objective function

$$f(\mathbf{x}) = \mathbf{c}^T \mathbf{x} \tag{5.1}$$

subject to a set of constraint equations

$$\mathbf{Ax} = \mathbf{b} \tag{5.2}$$

All n inequality constraint equations must be transformed into equality equations by including unknown positive slack variables. Thus, the size of the design variables vector $\mathbf{x}$ will be the number m of real design variables we wish to determine in order to minimize f plus the number n of slack variables. In consequence, matrix $\mathbf{A}$ is $n \times m$, the "available resources" vector $\mathbf{b}$ will be $n \times 1$, and vector $\mathbf{c}$ will be $(n + m) \times 1$, the same size of vector $\mathbf{x}$.

We strongly recommend to write a main program MATLAB script where we may first furnish a set of options that may differ from the default setting of this function such as, typically,

$$\text{options} = \text{optimset}('\text{LargeScale}', '\text{off}', '\text{TolCon}', 1e-8, '\text{TolX}', 1e-8)$$

Next, we may provide lower and upper bound for the design variables. In linear programing problems it is usual to have zero lower bound values and no upper bound values specification.

We may also provide an initial design, a line matrix with initial values of the design variables. Again, in linear programming it is not usual to do so.

The fundamental feature of this main program script is the optimization function call:

[x, FunVal, ExitFlag, Output] = linprog(c, [ ], [ ], A, b, Lb, Ub, x0, options)

Any information not used, such as the absence of the upper bound and initial values, in a particular problem, must be substituted by an empty matrix [ ].

5.2 Examples

5.2.1 Nonlinear cable problem

Consider the steel cable (Young's modulus $E = 21{,}000\,\text{kN/cm}^2$) originally straight under prestress force $N_0 = 50\,\text{kN}$ of Figure 5.1. At a point at a distance L_a from the left support and L_b from the right support, a horizontal force $H = 100\,\text{kN}$ and a vertical force $V = 200\,\text{KN}$ are applied. The design variables are u, the horizontal displacement, renamed x_1 (in cm), and the vertical v, renamed x_2 (in cm).

It is important to note that this is a nonlinear problem, in which equilibrium can only be written in the deformed position, initially unknown.

The basic idea is that stable equilibrium corresponds to a minimum of the total potential energy:

$$\Pi = U - W$$

where U is the strain energy and W is the work of external forces

$$W = Hu + Vv$$

The change in length is

$$\Delta_a = L'_a - L_a = \sqrt{v^2 + (L_a + u)^2} - L_a$$

$$\Delta_b = L'_b - L_b = \sqrt{v^2 + (L_b - u)^2} - L_b$$

The axial forces are

$$N_a = k_a \Delta_a$$

$$N_b = k_b \Delta_b$$

where

$$k_a = \frac{EA}{L_a}, \quad k_b = \frac{EA}{L_b}$$

Thus, the strain energy is

$$U = (N_0 + N_a/2)\Delta_a + (N_0 + N_b/2)\Delta_b$$

A steel cable with a length of 200 cm and transverse section $A = 1\,\mathrm{cm}^2$ was adopted, divided into two equal lengths.

Using Matlab's fminunc function in the solution, for unconstrained minimization, results $x_1 = 0.248$ cm, $x_2 = 20.6733$ cm, and the total potential energy at this minimum point is -3 kJ. As an exercise, check the equilibrium of the central node.

The correspondent MATLAB script is provided in the Appendix.

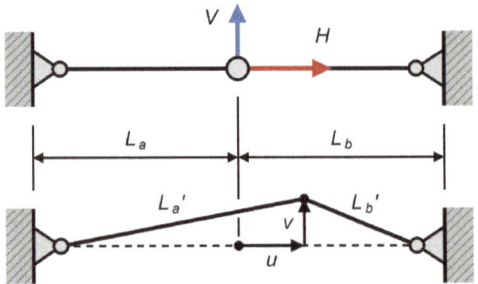

Figure 5.1: Cable in original position and in displaced position.

5.2.2 Eccentrically loaded tubular column

Minimize the mass of a tubular steel column of Figure 5.2, clamped in the base and free at the upper end where an eccentric compression load is applied.

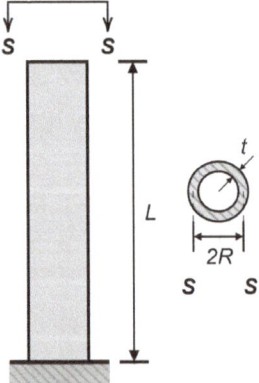

Figure 5.2: Column to be optimized.

Data:

P	Vertical compressive load, 100 kN
e	Eccentricity of load application, 2% of section average radius
L	Column length, 5 m
R	Section average radius, m
t	Steel wall thickness, m
E	Elastic modulus, 210 GPa
$\bar{\sigma}$	Allowable normal stress, 250 MPa
Δ	Allowable displacement, 0.25 m
ρ	Density, 7,850 kg/m³
A	Transverse section area, $2\pi R t$, m²
I	Transverse section moment of inertia, $\pi R^3 t$, m⁴
W	Bending modulus, $\pi R^2 t$, m³

From the resistance of materials, the following design expressions are taken.
Normal stress:

$$\sigma = \frac{P}{A}\left[1 + \frac{eA}{W}\sec\left(L\sqrt{\frac{P}{EI}}\right)\right] \leq \bar{\sigma}$$

Critical buckling load:

$$P_{cr} = \frac{\pi^2 EI}{4L^2} \geq P$$

Lateral displacement:

$$\delta = e\left[\sec\left(L\sqrt{\frac{P}{EI}}\right) - 1\right] \leq \Delta$$

In addition to these constraints given by the resistance of materials, there are design conditions that

$$\frac{R}{t} \leq 50 \qquad \text{and} \qquad 0.01 \leq R \leq 1, \quad 0.005 \leq t \leq 0.2$$

The design variables are, of course, the section radius (×1) and the steel wall thickness (×2). The objective function to be minimized is the mass of the column ρLA.

Answer: $R = 0.0537$ m; $t = 0.0050$ m

The correspondent MATLAB script is provided in the Appendix.

5.2.3 Constrained optimization of a statically loaded redundant truss

Consider the 3-time redundant truss of Figure 5.3. The cross-sectional areas are x_1 (in m²) for vertical and horizontal bars, of unit length, and x_2 (in m²) for diagonal bars, of length $\sqrt{2}$ m.

It is important to remember that in a redundant (hyper-static) structure the distribution of internal forces depends on the dimensions of the cross sections.

It is assumed a gravity load $P = 1$ kN applied to the mobile pinned support 2. The vertical displacements $p1$ and $p2$ of these mobile pinned supports are the only unknowns in the solution by the displacement method (or stiffness method, or equilibrium method).

The properties of the material, supposed to be homogeneous, isotropic, and linear elastic, are: allowable stress 10 kN/m²; elastic modulus $E = 100$ kN/m²; density $\rho = 1,000$ kg/m³.

The objective function to be minimized is the total mass of the truss:

$$f = \rho \left(3\, x_1 + 2\, \sqrt{2}\, x_2 \right)$$

subjected to the inequality constraint: normal stress less than the allowable stress of the material, buckling neglected.

The determination of the two unknown displacements, using the displacement method, is obtained from the algebraic linear system solution:

$$[K]\{p\} = \{P\}$$

$$E \begin{bmatrix} x_1 + x_2\, \sqrt{2}/4 & -x_1 \\ -x_1 & x_1 + x_2\, \sqrt{2}/4 \end{bmatrix} \begin{Bmatrix} p_1 \\ p_2 \end{Bmatrix} = \begin{Bmatrix} P_1 \\ P_2 \end{Bmatrix} = \begin{Bmatrix} 0 \\ -P \end{Bmatrix}$$

The normal forces on horizontal bars 2 and 3 are zero.

The normal tensile force on vertical bar 1 is

$$N_1 = E x_1\, (p_1 - p_2)$$

The normal tensile force on diagonal bar 4 is

$$N_4 = -E x_2 p_2/2$$

The normal compression force on diagonal bar 5 is

$$N_5 = E x_2 p_1/2$$

Matlab's fmincon function solution: $x1 = 0.0333$ m², $x2 = 0.0943$ m², minimum total mass of the truss 366.66 kg.

The correspondent MATLAB script is provided in the Appendix.

5.2.4 Frequency optimization of a redundant truss

Consider the 3-time redundant truss of Figure 5.4. The cross-sectional areas are x_1 (in m²) for vertical and horizontal bars, of unit length, and x_2 (in m²) for diagonal bars, of length $\sqrt{2}$ m. An undamped free vibration frequency bound is required.

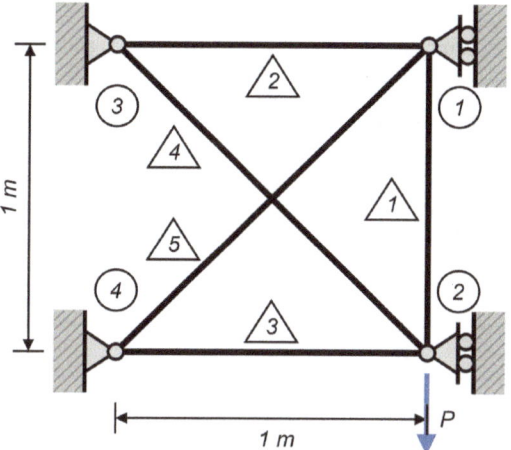

Figure 5.3: Sample redundant truss.

It is important to remember that in a redundant (hyper-static) structure the distribution of internal forces depends on the dimensions of the cross sections.

The vertical displacements p_1 and p_2 of the mobile pinned supports are the only two generalized coordinates of the problem.

The properties of the material, supposed to be homogeneous, isotropic, and linear elastic, are: elastic modulus $E = 100$ kN/m²; density $\rho = 1{,}000$ kg/m³.

The objective function to be minimized is the total mass of the truss:

$$f = \rho \left(3\, x_1 + 2\, \sqrt{2}\, x_2 \right)$$

subjected to the inequality constraints: the first frequency f_1 must be larger than 1 Hz and the second one f_2 less than 2 Hz. These two undamped free vibrations frequencies are obtained from the eigenvalue problem

$$\left[[K] - \omega^2 [M] \right] \{\varphi\} = \{0\}$$

where the stiffness matrix is

$$[K] = E \begin{bmatrix} x_1 + x_2\ \sqrt{2}/4 & -x_1 \\ -x_1 & x_1 + x_2\ \sqrt{2}/4 \end{bmatrix}$$

and a simplified lumped mass matrix is adopted as

$$[M] = \rho \begin{bmatrix} x_1 + x_2\ \sqrt{2}/2 & 0 \\ 0 & x_1 + x_2\ \sqrt{2}/2 \end{bmatrix}$$

Matlab's function eig(K,M) returns the squares of the two frequencies, in rad/s. Matlab's fmincon function solution: $x_1 = 0.01$ m^2, $x_2 = 0.0531$ m^2, minimum total truss mass 180 kg.

The correspondent MATLAB script is provided in the Appendix.

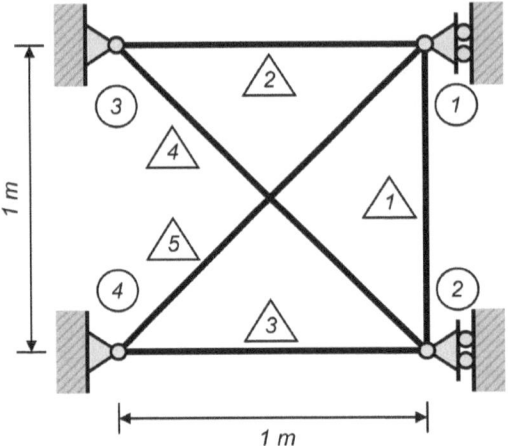

Figure 5.4: Sample redundant truss with frequency constraint.

5.2.5 Thickness optimization of a rectangular steel plate simply supported under uniformly distributed loading and its own weight

Optimize the thickness t (with a minimum of 10 mm) of a rectangular plate simply supported under uniformly distributed loading q given and its own weight. The smallest dimension, a, is in the x direction and the largest dimension, b, in the y direction.

Only b/a ratio between 1 and 2 is considered, rounded to the first decimal place.

The calculations are based on Table 8 of the book *Theory of Plates and Shells*, by Timoshenko, incorporated into the Matlab program.

Steel data: $E = 210e9$ N/m^2, Poisson's ratio 0.3, allowable stress 15e7 N/m^2, density 7,850 kg/m^3.

A maximum vertical displacement limit is also imposed in the middle of the span, equals to $a/400$.

For a 4 m square plate with a load of 1 tf/m^2 plus its own weight, a minimum thickness of 41.5 mm was obtained using the Matlab fmincon function, corresponding to a total plate mass of 5,217 kg.

The correspondent MATLAB script is provided in the Appendix.

5.2.6 Redundant wood planar portal frame

Optimize the pinned redundant wood planar portal frame of Figure 5.5. Column height $h = 3$ m and beam span $L = 6$ m.

Data: section width $b = 7.5$ cm, minimum section height 15 cm, characteristic force P_k applied in the middle of the beam 10 kN, characteristic strength of wood $f_{ck} = 40$ MPa, average elasticity modulus $E = 19{,}500$ MPa.

As a pinned-pinned planar portal frame is once hyperstatic, the stresses in the structure depend on the sections. The horizontal reactions in the pinned supports fixed at the base of the columns, X, were chosen as hyperstatic unknowns. Their value is:

$$X = -\frac{PhL^2/8I_b}{2h^3/3I_c + Lh^2/I_b}$$

where I_b and I_c are, respectively, the section moments of inertia of the beam and the column.

Thus, the column's maximum axial force and the bending moment are

$$N_c = \frac{P}{2} \quad M_c = Xh$$

and the beam's maximum axial force and bending moment are

$$N_b = X \quad M_b = \frac{PL}{4} - Xh$$

(neglecting self-weight).

Verification of wood parts to design axial force N_d and bending moment M_d is carried out using a **load and resistance factor design** scheme, as follows.

Design action: $P_d = \gamma_F P_k$, where it is adopted a load factor $\gamma_F = 1.5$.a

Design resistance: $f_{cd} = K_{mod} \frac{f_{ck}}{\gamma_w}$, where it is adopted resistance factor $\gamma_w = 1.4$,

$K_{mod} = 0.7$ is a modification factor due to wood particular behavior.

Accidental eccentricity: $e_a = \frac{L_0}{300}$, where L_0 is the bar length.

Initial eccentricity: $e_i = \frac{M_{d1}}{N_d}$, where M_{d1} is initial design bending moment.

Design eccentricity: $e_d = (e_i + e_a)\left(\frac{F_E}{F_E - N_d}\right)$, where $F_E = \frac{\pi^2 EI}{L_0^2}$ is Euler's buckling load.

Design bending moment: $M_d = N_d e_d$.

Verification: $\left(\frac{N_d}{A} + \frac{M_d}{W}\right)/f_{cd} \leq 1$,

where A and W are the section area and the elastic bending modulus.

Using Matlab's fmincon function, we arrive at a section of 7.5 × 15 cm for the columns and 7.5 × 27.5 cm for the beam. The minimum total mass of the frame is 191.4 kg.

The correspondent MATLAB script is provided in the Appendix.

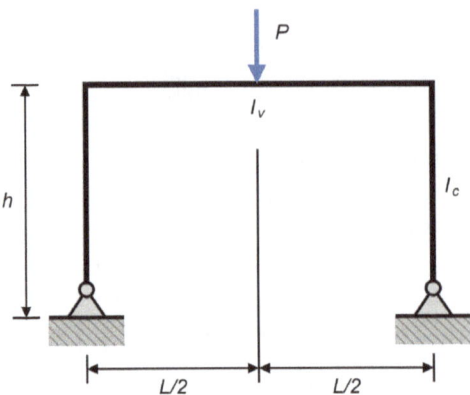

Figure 5.5: Reduntant wood portal frame.

5.3 Linear programming

We revisit Example 1.4 of Chapter 1, the same as Example 3.1 of Chapter 3 and Example 4.1 of Chapter 4, of maximization of profit of a toy factory (see Chapter 4):

$$f = -400x_1 - 600x_2$$

This is an example of a problem with linear constraint equations. These are:

$$x_1 + x_2 + x_3 = 16$$

$$x_1/28 + x_2/14 + x_4 = 1$$

$$x_1/14 + x_2/24 + x_5 = 1$$

where x_1 and x_2 are, respectively, the number of type A and type B toys produced. The other 3 design variables are slack variables, that is, surplus resources. In this case, related, respectively, to delivery, production and sales capabilities.

We call linprog MATLAB function

$$[x,FunVal,ExitFlag,Output] = linprog(c,[\ ],[\ ],A,b,Lb,[\ ],[\ ],options)$$

where

$$[c] = [\ -400 \quad -600 \quad 0 \quad 0 \quad 0\ ]$$

$$[A] = \begin{bmatrix} 1 & 1 & 1 & 0 & 0 \\ 1/28 & 1/14 & 0 & 1 & 0 \\ 1/14 & 1/24 & 0 & 0 & 1 \end{bmatrix}$$

$$[b] = [16 \quad 1 \quad 1]$$

We get the same results as those of the previous methods, that is:

$$[x] = [\, x_1 \quad x_2 \quad x_3 \quad x_4 \quad x_5 \,] = [\, 4 \quad 12 \quad 0 \quad 0 \quad 3/14 \,]$$

x_1 and x_2 are, respectively, the number of type A and type B toys that must be produced for the maximum profit objective function value $f = 8{,}800$ Euros.

The slack variables, that is, surplus resources, indicate that delivery and production departments are fully occupied and there is a little surplus capacity in the sales department.

The correspondent MATLAB script is provided in the Appendix.

Chapter 6
Numerical methods for nonlinear non-constrained optimization

When the optimization problem has a nonlinear objective function and/or nonlinear constraint functions, the problem is referred to as **nonlinear programming**.

There are three types of numerical methods for solving nonlinear programming problems:
- based on gradients;
- direct search;
- nature-inspired methods.

In this chapter we will deal only with numerical methods for nonlinear non-constrained optimization problems.

6.1 One-dimensional optimization: the golden section search

This section discusses the numerical solution to the problem of finding maximums or minimums of a single variable scalar function, without constraints. From the start, this function will be considered nonlinear, since the linear case in one dimension has no finite maximums or minimums.

One of the most used numerical methods for this purpose is the search using the golden section search. The golden ratio was first defined by Euclid (around 300 BC), in the construction of the pentagram, and stated as: "a straight line is said to be divided into extreme and mean ratio when the whole line is for the largest segment so as the latter is for the smaller." Let L_1 and L_2 be the largest and the smallest segments, respectively. Making

$$\varphi = \frac{L_1}{L_2} \tag{6.1}$$

according to Euclid's statement,

$$\varphi = \frac{L_1 + L_2}{L_1} \tag{6.2}$$

resulting in the second-order algebraic equation

$$\varphi^2 - \varphi - 1 = 0 \tag{6.3}$$

whose positive root is the so-called golden ratio

https://doi.org/10.1515/9783110625622-006

$$\varphi = \frac{1+\sqrt{5}}{2} = 1.618034\ldots \tag{6.4}$$

This value, sometimes known as *divine*, appears in a large number of relationships found in nature, science, and the arts. The reader is invited to do his own research on this interesting recurrence.

What is wanted is to determine the minimum of a one-dimensional function by a process in steps. It starts by defining an interval x_L and x_U within which the minimum is sought. Two points within this range are required to detect the occurrence of a minimum, and they will be chosen according to the golden ratio,

$$x_1 = x_L + d \quad \text{and} \quad x_2 = x_U - d, \qquad d = (\varphi - 1)(x_U - x_L) \tag{6.5}$$

The value of the function is calculated at these two interior points. Two results can occur.

1. If $f(x_1) < f(x_2)$, then $f(x_1)$ is the minimum of that range, and the domain to the left of x_2, from x_L to x_2, can be eliminated from the search because it does not contain the minimum. In this case, x_2 is the new x_L for the next iteration.
2. If $f(x_2) < f(x_1)$, then $f(x_2)$ is the minimum of that range, and the domain to the right of x_1, from x_1 to x_U, can be eliminated from the search because it does not contain the minimum. In this case, x_1 is the new x_U for the next iteration.

As the values of x_1 and x_2 were chosen using the golden ratio, it is not necessary to calculate all values of the function in the next iteration. For example, if hypothesis 1 above occurs, the former x_1 is the new x_2, and, thus, a new $f(x_2)$ does not need to be calculated, as it is the same as the old $f(x_1)$. To complete the algorithm, one determines the new x_1, applying Eq. (6.5), based on the new values of x_L and x_U. In case 2, the procedure is analogous.

It is shown that, in each iteration, the interval is reduced by a percentage of about 61.8%. So, for example, after 10 iterations, the range has decreased by about 0.618^{10}, 0.8% of its original length.

If it is desired to find the maximum of a function $f(x)$, instead of the minimum, just determine the minimum of that function with a changed sign, $F(x) = -f(x)$.

Example
Determine the minimum of function

$$f(x) = \frac{x^2}{10} - 2 \sin x$$

in interval $x_L = 0$ to $x_U = 4$.
   First iteration

$$d = 80.61803(4 - 0) = 2.4721$$
$$x_1 = 0 + 2.4721 = 2.4721 \qquad x_2 = 4 - 2.4721 = 1.5279$$

$$f(x_2) = -1.7647 \qquad f(x_1) = -0.63, \quad \therefore \quad f(x_2) < f(x_1)$$

The current minimum is $f(x_1) = -1.7647$.

Second iteration

$$d = 0.61803(2.4721 - 0) = 1.5279$$

$$x_1 = 1.5279 \qquad x_2 = 2.4721 - 1.5279 = 0.9443$$

$$f(x_2) = -1.5310, \quad \therefore \quad f(x_2) > f(x_1)$$

The current minimum is still $f(x_1) = -1.7647$.

After eight iterations, the location of the minimum is approximated by $x = 1.4427$, and the function value is −1.7755.

The search for the minimum using the golden ratio can be a routine that is repeated within a larger algorithm, as will be seen in Section 9.2.

6.2 Multidimensional optimization

In a multidimensional problem, that is, with several design variables, without constraints, formally:

find $\mathbf{x}^*$ to minimize $f(\mathbf{x})$.

In each step k, vector $\mathbf{x}^{(k)}$ is known and one must project a new value $\mathbf{x}^{(k+1)}$ by

$$\mathbf{x}^{(k+1)} = \mathbf{x}^{(k)} + \Delta\mathbf{x} \tag{6.6}$$

where

$$\Delta\mathbf{x} = \alpha_k \mathbf{d}^{(k)} \tag{6.7}$$

$\mathbf{d}^{(k)}$ is the desired direction and α_k is the length of the step in that direction. The algorithm is

Algorithm
1. Estimate a reasonable initial design $\mathbf{x}^{(0)}$, $k = 0$,
2. compute direction $\mathbf{d}^{(k)}$ at point $\mathbf{x}^{(k)}$,
3. check convergency; if reached, stop, else
4. compute a positive step length α_k in direction $\mathbf{d}^{(k)}$, and $\mathbf{x}^{(k+1)} = \mathbf{x}^{(k)} + \Delta\mathbf{x}$,
5. $k = k + 1$ and go to step 2.

To minimize the objective function, at each step its value must decrease

$$f\left(\mathbf{x}^{(k+1)}\right) < f\left(\mathbf{x}^{(k)}\right) \tag{6.8}$$

$$f\left(\mathbf{x}^{(k)}\right) + \alpha_k \mathbf{d}^{(k)} < f\left(\mathbf{x}^{(k)}\right) \tag{6.9}$$

Approaching the left side in Taylor's series about point $\mathbf{x}^{(k)}$

$$f\left(\mathbf{x}^{(k)}\right) + \alpha_k\left(\mathbf{c}^{(k)} \cdot \mathbf{d}^{(k)}\right) < f\left(\mathbf{x}^{(k)}\right) \tag{6.10}$$

where $\mathbf{c}^{(k)}$ is the gradient of the objective function at that point, and the dot product between the vectors in parentheses must be negative,

$$\left(\mathbf{c}^{(k)} \cdot \mathbf{d}^{(k)}\right) < 0 \tag{6.11}$$

as α_k was defined as always positive. This implies that the angle between these two vectors, the gradient and the desired direction, must be between 90° and 270°.

Any vector $\mathbf{d}^{(k)}$ satisfying this inequality, the condition of descent of the objective function, is a desirable direction. Methods of this type are descent methods "downhill," in which one wants to reach the bottom of the valley of the objective function from a higher point.

The determination of the step length is a one-dimensional or inline search

$$\mathbf{x}^{(k+1)} = \mathbf{x}^{(k)} + \alpha \mathbf{d}^{(k)} \tag{6.12}$$

$$f\left(\mathbf{x}^{(k+1)}\right) = f\left(\mathbf{x}^{(k)}\right) + \alpha \mathbf{d}^{(k)} = \bar{f}(\alpha) \tag{6.13}$$

$\bar{f}(\alpha)$ is a function of only α. When $\alpha = 0 \rightarrow \bar{f}(0) = f(\mathbf{x}^{(k)})$. Therefore, in order to decrease the value of the objective function, as desired, $\bar{f}(\alpha)$ should always decrease. This is a one-dimensional minimization problem without constraints. The necessary condition for a minimum is

$$\frac{d\bar{f}}{d\alpha} = 0 \tag{6.14}$$

and the sufficient condition is

$$\frac{d^2\bar{f}}{d\alpha^2} > 0 \tag{6.15}$$

Simplifying the notation by deleting the slash over the function, and applying the chain rule,

$$\frac{df\left(\mathbf{x}^{(k+1)}\right)}{d\alpha} = \frac{\partial f^T\left(\mathbf{x}^{(k+1)}\right)}{\partial \mathbf{x}} \frac{d\left(\mathbf{x}^{(k+1)}\right)}{d\alpha} = \nabla f\left(\mathbf{x}^{(k+1)}\right) \cdot \mathbf{d}^{(k)} = \mathbf{c}^{(k+1)} \cdot \mathbf{d}^{(k)} = 0 \tag{6.16}$$

That is, that the desired direction is orthogonal to the gradient vector in each step.

6.3 Gradient method: the steepest decent

Algorithm

1. Estimate a reasonable initial design $\mathbf{x}^{(0)}$, make $k = 0$, adopt a small convergency parameter $\varepsilon > 0$,
2. compute the gradient of the objective function at the current point $\mathbf{x}^{(k)}$, $\mathbf{c}^{(k)} = \nabla f(\mathbf{x}^{(k)})$,
3. compute length of the gradient vector, $\|\mathbf{c}^{(k)}\|$; if $\|\mathbf{c}^{(k)}\| < \varepsilon$, stop, as $\mathbf{x}^{(k)} = \mathbf{x}^*$, else, continue,
4. make search direction in this step $\mathbf{d}^{(k)} = -\mathbf{c}^{(k)}$,
5. compute the step size at this point, minimizing

$$f(\alpha) = f(\mathbf{x}^{(k)}) + \alpha\mathbf{d}^{(k)},$$

using the Golden section, for example,
6. update design $\mathbf{x}^{(k+1)} = \mathbf{x}^{(k)} + \alpha\mathbf{d}^{(k)}$, make $k = k + 1$ and return to step 2.

Example 1

Minimize

$$f(\mathbf{x}) = x_1^2 + x_2^2 - 2x_1x_2$$

adopted initial design $\mathbf{x}^{(0)} = [1 \quad 0]^T$, $k = 0$ and $\varepsilon = 10^{-4}$,

$$\mathbf{c}^{(0)} = [2x_1 - 2x_2 \quad 2x_2 - 2x_1]^T = [2 \quad -2]^T$$

$$\|\mathbf{c}^{(0)}\| = 2\sqrt{2} > \varepsilon, \text{ continue,}$$

$$\mathbf{d}^{(0)} = -\mathbf{c}^{(0)} = [-2 \quad 2]^T$$

minimize $f(\alpha) = f\left(\mathbf{x}^{(0)}\right) + \alpha\mathbf{d}^{(0)}$, where $\mathbf{x}^{(0)} + \alpha\mathbf{d}^{(0)} = [1 - 2\alpha \quad 2\alpha]^T$

$$f(\alpha) = f\left(\mathbf{x}^{(0)}\right) + \alpha\mathbf{d}^{(0)} = (1 - 2\alpha)^2 + (2\alpha)^2 - 2(1 - 2\alpha)(2\alpha)$$

$$f(\alpha) = 16\alpha^2 - 8\alpha + 1$$

$$\frac{df}{d\alpha} = 0, \quad \therefore \quad \alpha_0 = 1/4$$

$$\frac{d^2f}{d\alpha^2} = 32 > 0$$

$$\mathbf{x}^{(1)} = \mathbf{x}^{(0)} + \alpha_0 \, \mathbf{d}^{(0)} = [0.5 \quad 0.5]^T$$

$$\mathbf{c}^{(1)} = [0 \quad 0]^T, \|\mathbf{c}^{(1)}\| < \varepsilon, \text{ stop.}$$

Answer: $\mathbf{x}^* = [0.5 \quad 0.5]^T$, and the minimum value of the objective function is zero at that point.

Example 2

Minimize

$$f(\mathbf{x}) = x_1^2 + x_2^2 + x_3^2$$

adopted initial design $\mathbf{x}^{(0)} = [1 \quad 1 \quad 2]^T$, $k = 0$ and $\varepsilon = 10^{-4}$,

$$\mathbf{c}^{(0)} = [2x_1 \quad 2x_2 \quad 2x_3]^T = [2 \quad 2 \quad 4]^T$$

$\|\mathbf{c}^{(0)}\| > \varepsilon$, continue,

$$\mathbf{d}^{(0)} = -\mathbf{c}^{(0)} = [-2 \quad -2 \quad -4]^T$$

minimize $f(\alpha) = f(\mathbf{x}^{(0)}) + \alpha \mathbf{d}^{(0)}$, where $\mathbf{x}^{(0)} + \alpha \mathbf{d}^{(0)} = [1 - 2\alpha \quad 1 - 2\alpha \quad 2 - 4\alpha]^T$

$$f(\alpha) = (\mathbf{x}^{(0)}) + \alpha \mathbf{d}^{(0)} = (1 - 2\alpha)^2 + (1 - 2\alpha)^2 + (2 - 4\alpha)^2$$

$$f(\alpha) = 24\alpha^2 - 24\alpha + 6$$

$$\frac{df}{d\alpha} = 0, \quad \therefore \quad \alpha_0 = 1/2$$

$$\frac{d^2 f}{d\alpha^2} > 0$$

$$\mathbf{x}^{(1)} = \mathbf{x}^{(0)} + \alpha_0 \, \mathbf{d}^{(0)} = [0 \quad 0 \quad 0]^T$$

$$\mathbf{c}^{(1)} = [0 \quad 0 \quad 0]^T, \mathbf{c}^{(1)} < \varepsilon, \text{ stop.}$$

Answer: $\mathbf{x}^* = [0 \quad 0 \quad 0]^T$, and the minimum value of the objective function is zero at that point.

Note: In these two examples, it was possible to determine the minimum of the function $f(\alpha)$ by the necessary and sufficient conditions. In more complex cases, it may be necessary to use a numerical method such as the golden section.

Chapter 7
Nature-inspired optimization methods

In this chapter, some methods of nature-inspired design optimization are presented. Nature, as it is known, produces, almost always, individuals optimized to the ambient conditions. Among these methods, the most researched in recent times are genetic algorithms and, as a result, they are the most developed and known. In this case, the problem normally deals with discrete variables. Several other methods of this family will have their bases exposed here, but not implemented.

7.1 Problem setting for discrete variables

The structural optimization problems for discrete variables solved in this work can be presented in the form:

determine $\mathbf{b} \in \mathfrak{R}^n$ that minimize as objective function

$$f(\mathbf{b}, T) = \bar{f}(\mathbf{b}, T) + \int_0^T \tilde{f}(\mathbf{b}, \mathbf{z}, \dot{\mathbf{z}}, \ddot{\mathbf{z}}, t)dt \tag{7.1}$$

Subjected to static constraints:

$$g_i = \bar{g}_i(\mathbf{b}, T) + \int_0^T \tilde{g}_i(\mathbf{b}, \mathbf{z}, \dot{\mathbf{z}}, \ddot{\mathbf{z}}, t)dt \begin{cases} = 0 \text{ for } i = 1, \ldots, l \\ \leq 0 \text{ for } i = l+1, \ldots, m \end{cases} \tag{7.2}$$

Subjected to dynamic constraints:

$$\tilde{g}_i = g_i(\boldsymbol{b}, \mathbf{z}, \dot{\mathbf{z}}, \ddot{\mathbf{z}}, t) \begin{cases} = 0 \text{ for } i = m+1, \ldots, l' \\ \leq 0 \text{ for } i = l'+1, \ldots, m' \end{cases} \text{ for } t \in [0, T] \tag{7.3}$$

with

$$b_i \in \mathbf{b}_i \equiv \{ b_{i1} \quad b_{i2} \quad \ldots \quad b_{iN_{Ei}} \}, \tag{7.4}$$

where $b_{i1}, b_{i2}, \ldots, b_{iNEi}$ are the N_{Ei} possible discrete values of variable b_i.

In these problems, state variables $\mathbf{z}(t)$, the displacement vector, must satisfy the equations of motion:

$$\mathbf{M}\ddot{\mathbf{z}} + \mathbf{C}\dot{\mathbf{z}} + \mathbf{K}\mathbf{z} = \mathbf{p}(t) \qquad \forall\, t \in [0, T] \tag{7.5}$$

with initial conditions $\mathbf{z}(0) = \mathbf{z}_0$ and $\dot{z}(0) = \dot{z}_0$. Constraints with dynamic responses are explicit functions of the state variables and implicit functions of the design variables.

https://doi.org/10.1515/9783110625622-007

To evaluate them, it is necessary to solve system (7.5) that needs to be integrated numerically. For this, the interval $[0, T]$ must be discretized.

7.2 Differential evolution algorithm

The differential evolution algorithm works with a population of designs. At each iteration, called a generation, a new design is generated using some current designs and random operations. If the new design is better than a pre-selected parent, then it takes its place in the population. Otherwise, the old design is preserved and the process is repeated.

Compared to genetic algorithms, they are easier to implement computationally, requiring no work with binary numbers. The basic steps are as follows.
1. Generation of the initial design population, in a large number Np. Each design, point or vector, is also called a chromosome, and its components are genes. To cover the space of the design variables, it is interesting to draw random values between the lower and upper limits of these variables.
2. Mutation to generate so-called donor design vectors. Three vectors of the current population are selected. The difference between two of these vectors is multiplied by a scale factor and added to the third vector for donor formation. In addition, a parent vector is selected.
3. Crossover/recombination to generate the so-called tentative design vectors. In this step, the donor vector exchanges some genes (components) with the parent vector (the crossing).
4. Selection, that is, acceptance or rejection of tentative design vectors using an adequacy function, which is usually the cost function (objective function).

7.3 Ants colony

This algorithm emulates the ant's food-seeking behavior. It is related to graphs representing the search for an optimal path between the colony and the food source. A similar classic problem is that of the traveling salesman who seeks the optimal route to visit all his clients on a trip. Graph theory was initiated by Euler in the famous problem of the Konigsberg bridges.

The method involves the biological concept of pheromone, derived from the Greek pherin (carry) and hormone (stimulate). It refers to the chemical substance secreted by insects that stimulates social behavior among members of the same species. Ants deposit this product on their way, leading others to follow it instead of another. The passing of many insects on the same trail increases the density of pheromone, indicating the most used or favorable paths. If a path is abandoned (for not taking the food source yet to explore), the compost evaporates over time.

7.4 Particles cloud

This algorithm simulates the social behavior of schools of fish or flocks of birds. As is well known, individuals from large groups of these animals seem to behave in exactly the same way, without humanly sensible communication between them, when searching for food or escaping from predators. Recent research seems to indicate that because all members of the group are under the same stimulus, they tend to respond to it in exactly the same way, in the limit, which would be the optimal response to that stimulus.

The method calls a particle or agent a given individual in the group, and its position corresponds to a potentially different design. He must memorize his current position as well as the best position reached until that stage, until all the particles reach the best possible position.

7.5 Genetic algorithms

The great English scholar Charles Darwin (1809–1882), in his maximum work *On the Origin of Species by Means of Natural Selection*, from 1859, bequeathed to mankind the theory that today is considered definitive about the evolution of species. In short, he says that as the generations pass, small changes in their characteristics can happen at random. Favorable changes tend to be passed on to the next generations by more adapted or more competitive parents, while unfavorable ones tend to disappear. In this way, species evolve, presumably for the better. This is called Natural Selection.

It is very important to note the role of the environment in the process. A mutation can lead to an individual better adapted or not to the ambient conditions. Changes, also random, in these conditions can change the individual's chances of survival. A classic example is that of white butterflies that were adapted to light trunks of trees in regions of Great Britain. Due to the industrial revolution that darkened these trunks with pollution, dark butterflies began to proliferate.

With the extraordinary advancement of computational capacity available today, a very interesting idea in numerical simulations in all branches of science is the application of Monte Carlo–type methods. In these, a large number of simulations can be performed with random variation of parameters. From the cloud of results obtained, approximate assessments, preferably based on statistics, can be made of the characteristics of a very complex phenomenon depending on a very large number of parameters, sometimes little known.

Based on these two ideas, natural selection and Monte Carlo methods, genetic algorithms appeared. In short, in them, a set of randomly generated designs goes from "generation" to "generation," with introduction of random small changes in their characteristics. The objective function is calculated at each stage for all designs in order to assess whether the change is favorable or not, according to some criterion, and whether it should be passed on to the next generation or not.

As one can see, it depends on a large processing capacity and there is no reliable or guaranteed way to affirm that a certain set of characteristics is the optimal design that is sought.

Important: genetic algorithms, at least in most current versions, do not take into account the random changes in the "environment" (the design conditions). This usually remains constant during the optimization process.

Here are some definitions:

Population. It is the set of points, each of them a different randomly generated design. N_p is the number of these designs or the size of the population.

Generation. It is an iteration of the algorithm.

Chromosome. It is the vector of values of the design variables of a given point. Also called a design or a genetic chain.

Gene. It is a particular value (a scalar) of one of the design variables, a component of the chromosome.

Fitness function. Defines the relative importance of a design. A higher value is a better design.

Reproduction. It is an operator in which an old design is passed on to a new generation according to their level of aptitude. It is the selection process.

Mating pool. Part of the population that will participate in the reproduction process, chosen from among the most apt members of it, evaluated by the aptitude function.

Mating. It is the process by which selected members of a new population exchange characteristics with each other.

Mutation. It is the random change of any characteristic (gene).

Stopping criterion. If the improvement in the best objective function is less than a given small value for the last consecutive iterations (in a defined number), or the number of iterations exceeds a specified value, the algorithm is terminated.

Immigration. Introduction of completely new designs in the population, in search of diversity, in some iterations when convergence is slow.

A genetic algorithm (GA) begins with a set of designs, called the initial or first-generation population. Each design, also called a member of the population, is represented by a binary string. Of this generation, the next is formed using three operators: reproduction, crossing, and mutation. Reproduction is an operator through which a current (current) design is introduced into a new population in such a way that its characteristics are transferred to the most suitable members of the population. Mating corresponds to allowing, at random, certain members of the population to exchange characteristics of the designs between them. The mutation operator is used to protect

the process from premature complete loss of valuable genetic material during reproduction and breeding. In the end, a design with a better aptitude is adopted as the
optimal design. It is then observed that the need arises in this algorithm to measure
the suitability of a design and also to define procedures for random selection of members of the population. The GA does not require functions in the process to be differentiable. It is only assumed that these functions can be computed for a given design.

The first task in a genetic algorithm is to represent the designs. In optimization
problems with discrete variables, each variable is associated with a vector that represents the discrete values that this variable can assume. Then a scheme needs to
be defined to select a value for each design variable. This scheme can be carried out
with the creation of a binary string B. To elucidate how this string is determined,
the example below was elaborated.

Example 7.1: Creation of a binary string B for a GA problem.
Consider a discrete optimization problem with two design variables: b_1 and b_2. Variable b_1 can as-
sume any value of the set represented by the vector $\mathbf{b}_1^t = [b_{11}\ b_{12}\ b_{13}\ b_{14}\ b_{15}\ b_{16}]$, while variable b_2
can assume any value of the set represented by the vector $\mathbf{b}_2^t = [b_{21}\ b_{22}\ b_{23}\ b_{24}\ b_{25}]$. To define a
design in this optimization problem, it is necessary to select a value for b_1 and another for b_2,
among values defined by $\mathbf{b}_1$ and $\mathbf{b}_2$. This choice is based on the numbering of the position of each
value that can be assumed by a given design variable, however the number that represents the position is given on the binary basis. See Table 7.1 to clarify the procedure:

Table 7.1: Example of creation of binary strings to represent a design variable.

Position	Variable			
	b_1		b_2	
	String	Value	String	Value
1	000	b_{11}	000	b_{21}
2	001	b_{12}	001	b_{22}
3	010	b_{13}	010	b_{23}
4	011	b_{14}	011	b_{24}
5	100	b_{15}	100 to 111	b_{25}
6	101 to 111	b_{16}	—	—

Here, the chosen value b_{13} for design variable b_1 is represented by string {010}, position
number 3 in vector $\mathbf{b}_1$, while any string from {101} to {111} represents value b_{16}. Simi-
larly, binary representation of value b_{21} for design variable b_2 is {000}, number 1 string
of vector $\mathbf{b}_2$. Note that the decimal numeric value of the binary string that represents
the position 1 is 0, of position 2 is 1, of position 3 is 2, etc. Thus, design $\mathbf{b}^t = [b_{13}\ b_{21}]$ is
represented by string $\mathbf{B} = \{010000\}$. Each component of $\mathbf{B}$ (0 or 1) is called a bit or gene,

while **B** is called the design DNA. This sample string has six bits. It is important to note that in the example above for position 6 of the variable b_1, for example, there is more than one binary string to represent it. This fact privileges this value to a certain extent for the design variable. Arora et al (1994) present a procedure to avoid this type of problem.

As a procedure for the determination of **B** (binary string) is defined, a first population with N_p members is created using N_p strings **B**. The size of the population is kept constant from one generation to the next. The next task is to define a breeding criterion. This criterion is based on an aptitude function. This is defined such that a design with a higher aptitude value has a greater probability of selection for reproduction. The most suitable members of the population are selected by the reproduction process and grouped in a mating pool from which members are selected for breeding and mutation. The next operator, crossover, is carried out between two designs called relatives. For this purpose, two designs are selected at random. These are called parent designs. Then the related designs are broken into segments (sets of consecutive bits) and some of these segments are exchanged with the corresponding ones of the other relative. The mutation arbitrarily changes the value of the gene (from 0 to 1 or vice versa) according to a predetermined probability.

The different GAs differ from one another according to the implementation of reproduction, crossover, and mutation. The GA used in this work is based on the one originally developed by Arora et al (1994) and later improved by Kocer and Arora (1999), which is available in the computational package IDESIGN of the *Optimal Design Laboratory* of the University of Iowa. In this implementation, non-viable designs (those that violate restrictions) are not rejected and violations of restrictions are used to define the following penalty function:

$$p_i = f_i + RK_{bi} \tag{7.6}$$

where f_i is the objective function for the i-th design, $R > 0$ is a penalty parameter and K_{bi} is the maximum constraint violation of the i-th design, that is, K_b for the i-th design. The penalty parameter must be chosen carefully, in such a way that neither f_i nor RK_{bi} dominate (7.6). In the implementation adopted in this work, R is calculated based on the values of the objective function for the first generation:

$$R = \frac{\sum_{i=1}^{N_p} f_i}{N_p \varepsilon} \tag{7.7}$$

where ε is the acceptable value for the maximum constraint violation.

Based on the definition of the penalty function (7.6), the aptitude function for the i-th design is then defined as

$$F_i = (1 + \vartheta)p_{\max} - p_i \tag{7.8}$$

where $\vartheta > 0$ is a small number used to force convergence and $p_{\max}$ is the highest value of the penalty function for the first generation. It is observed that the value of $p_{\max}$ is constant for the rest of the design.

In the implementation of the mating, two designs are selected at random. Then a random number from the interval [0, 1] is generated. If this number is less than the probability of crossing P_c, then the crossing is performed; that is, the values of two consecutive bits are exchanged for the pair. Otherwise, these two relatives are neglected and a new pair is selected. The probability of crossing P_c adopted here is equal to 0.5.

Mutation is implemented in each bit of string **B**. This means that for each bit of the entire population a random number is selected, and if this number is greater than the probability of mutation P_m, mutation is performed. However, this implementation requires the selection of a large number of random numbers. This amount is equal to the size of the population multiplied by the number of bits that a design represents. With this, instead of selecting a random number for each bit, the expected number of mutations is calculated and then the various mutations are performed. This number is the same as $P_m N_p N_b$, where N_b is the number of bits that represents a design. In Example 7.1 we have $N_b = 6$. Thus, a random design is chosen. In this design a random bit is selected and, if its value is equal to 0, it will be changed to 1, and vice versa. This procedure will be repeated $P_m N_p N_b$ times. The mutation probability P_m adopted here is equal 0,3.

It is still necessary to determine the size of the population N_p, a stopping criterion, and a limit number for the number of iterations. The possible number of designs for a given problem is calculated as:

$$N_E = \prod_{i=1}^{n} N_{Ei} \tag{7.9}$$

where $N_{E\,i}$ is the number of discrete values for the ith variable and n the total number of design variables. In this example, $N_{E1} = 6$ and $N_{E2} = 5$ so that $N_E = 30$. One must also define ratio

$$\chi = \frac{N_E}{n_1} \tag{7.10}$$

The following procedure determines the value of N_p:

if $N_E \leq n_2\, n$ then
$$N_p = N_E$$
else
if $\chi < n_3\, n$ then
$$N_p = n_3\, n$$
else
if $\chi > n_4\, n$ then
$$N_p = n_4\, n$$
else
$$N_p = \chi$$

In the current implementation it is adopted $n_1 = 1,000$, $n_2 = 6$, $n_3 = 2$, and $n_4 = 8$.

One of the stopping criteria for the present algorithm is based on changing the values of the aptitude function and can be represented by Eq. (7.11):

$$\frac{\left[\max_{1 \le i \le N_p} p_i\right]_{k+1} - \left[\max_{1 \le i \le N_p} p_i\right]_{k}}{\left[\max_{1 \le i \le N_p} p_i\right]_{k=1}} \le \varepsilon' \tag{7.11}$$

where k is the kth iteration and ε' an adopted small number chosen to be 10^{-3}. According to Kocer and Arora (1999), a reasonable value for the maximum number of iterations would be $3\,n$. So the second stopping criterion adopted in this work is

$$k < 3n \tag{7.12}$$

Other stopping criteria can be obtained from Arora et al (1994).

Genetic algorithm:
Step 1 – Define a binary string **B** to represent a given design.
Step 2 – Generate N_p random strings (members of the population). Make $k = 0$.
Step 3 – Define penalty functions (7.6) and aptitude functions (7.8).
Step 4 – Compute aptitude values for all designs. Make $k = k + 1$.
Step 5 – Reproduction:
    5.1 Select a leader (a design) from the previous generation. Save this design twice. Copy one to the next generation and send the other to the mating pool.
    5.2 Calculate the probability of selecting each design using the equation

$$P_i = \frac{F_i}{\sum_{j=1}^{N_p} F_j}$$

and assemble the Figure 7.1 scheme

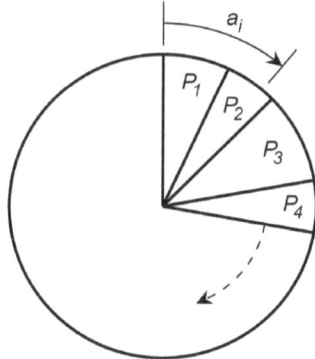

Figure 7.1: Roulette.

where the total area of the circle above is equal to 1.

5.3 Select at random, as shown above, $N_p - 1$ designs generating $N_p - 1$ random numbers a_i between [0, 1]. As shown above, for number a_i randomly generated the design was chosen with the probability P_3. These $N_p - 1$ designs and the leader form the mating group for the crossing and mutation operations. The same design can be chosen more than once while others may not be chosen at all.

Step 6 – Mating:

6.1 Select two designs (one pair for breeding) from the mating pool.

6.2 Generate a random number. If it is less than the mating probability, P_c, do the mating: select two consecutive bits in the string that represents one of the designs and change for the corresponding bits of the other design (cross the bits, those that belonged to one design will become the other). Go to 6.1 and repeat the process until all of the population's designs have been selected at least once.

Step 7 – Mutation:

7.1 Calculate the possible number of mutations, $N_m = P_m N_p N_b$.

7.2 Choose N_m designs from the mating pool. For each design, select a position in the string and change 0 to 1, or vice versa.

Step 8 – If the stopping criteria are met, that is, if Eqs. (7.11) and (7.12) are met, then stop the process, otherwise go to Step 4.

Example 7.2

Determination of the optimum profile for a truss bar subjected to traction. Latticed structures are quite common in engineering. See Figure 7.2 for an example of a telecommunication tower, latticed and in angle profiles. The main load are the forces due to the wind. The bars work mainly with traction and compression.

Figure 7.2: Typical telecommunication tower.

The main load is the loads due to the wind. The bars work mainly with traction or compression, depending on the wind direction, as shown in Figure 7.3.

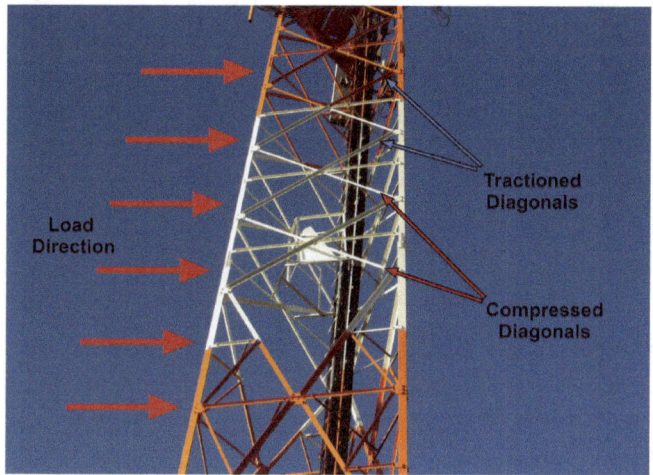

Figure 7.3: Scheme of loads and behavior of diagonals in relation to internal efforts.

The bars must meet the allowable stress limit; that is, the stress acting on the part must not be higher than the allowable value. In Figure 7.4 is shown an angle section subjected to a traction load $P = 10$ tf. Material is steel with allowable stress $\sigma_a = 2.25$ tf/cm^2. The cross-sectional area A is the design variable and the objective function of the optimization problem.

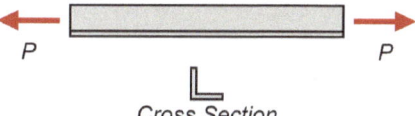

Cross Section

Figure 7.4: Angle member under traction.

The optimization problem is written as:

Determine $b_1 = A$ that minimizes $f(b_1) = b_1$, subject to constraint

$$g1 = P/A - \sigma_a \leq 0 \tag{7.13}$$

Design variable b_1 must assume one of the values shown in Table 7.2.

Table 7.2: Values that can be assumed for the design of the cross section.

Position	Variable		
	b_1		
	String	Angle	Area (cm²)
1	00	L 2″ × 1/8″	3.13
2	01	L 2″ × 3/16″	4.62
3	10	L 2″ × 1/4″	6.04
4	11	L 2″ × 5/16″	7.43

Beforehand, it can be concluded that design 01 is the one with the smallest area and meets the constraint (7.13).

Solving the problem using the genetic algorithm calculate $N_p = 4$, and the initial population shown in Table 7.3 is generated. Observe in this table that the design with the best aptitude is precisely the one related to string 01. In Figure 7.4 the roulette for the initial population is shown in Figure 7.5. Designs 2, 3, and 4 are the ones with the highest probability of reproduction. Design 1, which violates constraint (7.13), has a very low probability of being chosen for reproduction.

The design chosen as the leader is the more apt, that is, 01. Table 7.4 shows the designs of the mating group determined as described in Step 5 of the genetic algorithm.

Table 7.4 shows that the designs related to strings 01 and 10 were chosen for the mating group, while the other designs were passed over. Now applying the crossing and the mutation, and maintaining the leader, we have the second generation ($k = 1$), shown in Table 7.5. Again, the design related to string 01 is the one with the greatest aptitude. Note that the stop condition (7.11) was satisfied, as the result of this equation is equal to 0. The design considered as optimal is the one with the greatest aptitude, that is, the string 01.

Table 7.3: Initial population ($k = 0$).

Position	Variable			Penalty function calculation					Aptitude function		P_i	Accumulated P_i
	b_1											
String	Angle	Area	f_i	f_i	R	Kb_i	p_i	p_{max}	F_i			
1	00	L 2″ × 1/8″	3.13	3.13	5,305	0.945	5,015.762	5,015.762	5.016	0.03%	0.03%	
2	01	L 2″ × 3/16″	4.62	4.62	5,305	0.000	4.620	5,015.762	5,016.158	33.33%	33.36%	
3	10	L 2″ × 1/4″	6.04	6.04	5,305	0.000	7.430	5,015.762	5,013.348	33.31%	66.68%	
4	11	L 2″ × 5/16″	7.43	7.43	5,305	0.000	6.040	5,015.762	5,014.738	33.32%	100.00%	

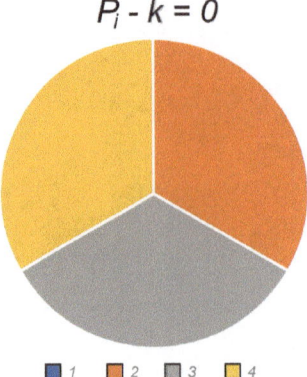

$P_i - k = 0$

■ *1* ■ *2* ■ *3* ■ *4*   **Figure 7.5:** Probability roulette.

The algorithm in this case converged in a single iteration and the optimal design was exactly the string 01, which is equivalent to a L 2″ × 3/16″ angle with a cross-sectional area of 4.62 cm². GA in this example was able to actually determine the optimal point of the system, but in general there is no guarantee that it will converge to the optimal value.

Table 7.4: Mating pool ($k = 1$).

Position	Variable b_1			Penalty function calculation					Aptitude function
	String	Angle	Area	f_i	R	Kb_i	p_i	p_{max}	F_i
1	01	L 2″ × 3/16″	4.62	4.62	5,305	0.000	4.620	5,015.762	5,016.158
2	01	L 2″ × 3/16″	4.62	4.62	5,305	0.000	4.620	5,015.762	5,016.158
3	10	L 2″ × 1/4″	6.04	6.04	5,305	0.000	6.04	5,015.762	5,014.738
4	10	L 2″ × 1/4″	6.04	6.04	5,305	0.000	6.040	5,015.762	5,014.738

Table 7.5: Second-generation population ($k = 1$).

Position	Variable			Penalty function calculation					Aptitude function			P_i	Accumulated P_i
		b_1											
	String	Angle	Area	f_i	R	Kb_i	p_i		p_{max}	F_i			
1	01	L 2″ × 3/16″	4.62	4.62	5,305	0.000	4.620		5,015.762	5,016.158		33.33%	33.33%
2	00	L 2″ × 1/8″	3.13	3.13	5,305	0.945	5,015.762		5,015.762	5.016		0.03%	33.37%
3	11	L 2″ × 5/16″	7.43	7.43	5,305	0.000	7.430		5,015.762	5,013.348		33.32%	66.68%
4	11	L 2″ × 5/16″	7.43	7.43	5,305	0.000	7.43		5,015.762	5,014.738		33.32%	100.00%

Chapter 8
Topology optimization and reliability

In this chapter, some methods of topology design optimization are presented. Topology optimization (TO) is a mathematical method which spatially optimizes the distribution of material within a defined domain, by fulfilling given constraints previously established and minimizing a predefined objective function.

As a related topic, reliability verification algorithms are also discussed.

8.1 Topology optimization

Optimization can be defined as a procedure by which it is possible to find a solution or a set of optimal solutions for a given function or set of functions, which govern a specific problem, subject to restrictions. So, in TO, there is the intention of providing the best distribution of material from a fixed design space. Hence, this innovation favors industries in different sectors, considering that designing mechanical parts and components with large stiffness and small weight has become a common necessity. It has been applied, for example, in aircraft wing spars webs, as represented in Figure 8.1.

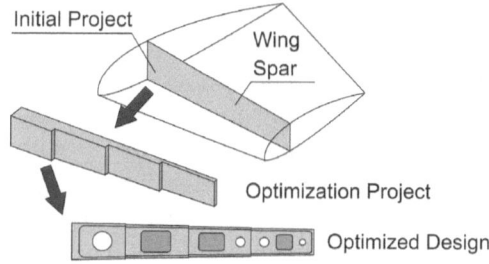

Figure 8.1: Process of topology optimization of an aircraft wing spar.

The TO procedure can be outlined as shown in Figure 8.2, which displays the use of the finite element method (FEM) together with an optimization algorithm, which introduces numerical strategies in the search for optimal engineering solutions, to obtain an optimized domain.

It is worth mentioning that in the practical application of TO, some aspects are fundamental. For example, in Figure 8.2 (obtained topology), points with intermediate shades between black and white, called gray scales, can appear. These points indicate the presence of elements with an intermediate thickness between the maximum and the null. These thicknesses may not be feasible to be implemented in practice, but they usually occur; that is, the presence of the gray scale is inherent in obtaining the optimal solution (Brasil, 2017). The image of the obtained structure by TO represents an excellent starting point that needs to be interpreted, in order to obtain the final design of the

https://doi.org/10.1515/9783110625622-008

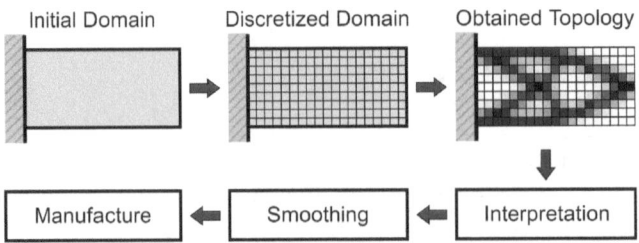

Figure 8.2: Topology optimization design procedure.

structure to be adopted in practice in the industry. This interpretation process is called refinement or smoothing, and can be done using image processing methods, or simply by designing a structure based on the image obtained by TO, often adhering to CAD/ CAE software. In some cases, the results generated by TO are not intuitive and it is necessary to check the final structure using the FEM. The last step is the manufacture of the structure.

8.2 The finite element method in topology optimization

To better exemplify this whole process, we will deal with some examples such as rectangular plates and beams with concentrated loads and distributed loads with various boundary conditions, using MATLAB, a high-level programming language that allows the solution of countless scientific problems with its accessible syntax, excellent debugging tools, and extensive graph manipulation tools. Therefore, it allows the user to focus on the physical and mathematical context of the optimization problem without being distracted by technical implementation problems. The optimization algorithm will consist of determining the thickness of each element, in order to minimize the total mass function of the structure, respecting the limits of allowable stresses imposed.

The FEM consists of discretizing the domain of the structure in several subdomains (bars, triangles, quadrilaterals, tetrahedrons, etc.), called elements, of small but finite dimensions, united in points called nodes. An equilibrium equation is assembled for each element, and then these equations are combined to determine an expression that represents the structure as a whole. Next, the displacements, strains, and stresses in the domain of the structure are determined.

The displacements of a point in a solid continuum are modeled by a vector **u**. In the case of a plate, in a two-dimensional domain of x and y axes, we have a 2×1 vector:

$$\mathbf{u} = \left\{ \begin{array}{l} u(x,y,t) \\ v(x,y,t) \end{array} \right\} \tag{8.1}$$

From the displacements, the strains are obtained by the application of a differential operator $\mathbf{D}$:

$$\varepsilon = \mathbf{D}\mathbf{u} \qquad (8.2)$$

so, in the case of a plate, we have the following 3×2 operator:

$$\varepsilon = \left\{ \begin{array}{c} \varepsilon_x \\ \varepsilon_y \\ \gamma_{xy} \end{array} \right\} = \mathbf{D}\mathbf{u} = \begin{bmatrix} \frac{\partial}{\partial x} & 0 \\ 0 & \frac{\partial}{\partial y} \\ \frac{\partial}{\partial y} & \frac{\partial}{\partial x} \end{bmatrix} \left\{ \begin{array}{c} u \\ v \end{array} \right\} \qquad (8.3)$$

The next step is to obtain the stress vector, from the strains, using, for simplicity, Hooke's law, in matrix form:

$$\sigma = \mathbf{E}\varepsilon = \mathbf{E}\mathbf{D}\mathbf{u} \qquad (8.4)$$

In the case of a plate, the stress vector is 3×1:

$$\sigma = \left\{ \begin{array}{c} \sigma_x \\ \sigma_y \\ \tau_{xy} \end{array} \right\} \qquad (8.5)$$

in the plane stress hypothesis, Hooke's law is expressed by the following 3×3 matrix:

$$\mathbf{E} = \frac{E_m}{1 - v^2} \begin{bmatrix} 1 & v & 0 \\ v & 1 & 0 \\ 0 & 0 & \frac{1-v}{2} \end{bmatrix} \qquad (8.6)$$

where v is the Poisson's ratio and E_m is Young's modulus.

Our sample structure will be a rectangular plate. In order to implement the FEM in MATLAB, it is necessary to input the dimensions of the structure to be analyzed and how it will be discretized. In this specific case, Argyris' four-node square elements will be used, and one must define the number of elements, nodes, and size for each element, in order to generate the coordinates of the nodes and the connectivity between them. Besides of that, we have to provide all the physical specificities of the material, such as Poisson's ratio and the Young's modulus.

```
%Dimensions
compr=8;
alt=2;
%divisions
ndx=24; ndy=6; %number of divisions on x and y
nel=ndx*ndy; %number of elements
nno=(ndx+1)*(ndy+1); %number of nodes
dx=compr/ndx; dy=alt/ndy; %size of each element
```

```
%admissible stress of steel
csadm = 225e6;
%generating the nodes coordinates
gcoord=zeros(nno,2);
x=0;y=0;k=0;
for i=1:ndx+1
  for j=1:ndy+1
    k=k+1;
    gcoord(k,1)=x;
    gcoord(k,2)=y;
    y=y+dy;
  end
  y=0;
  x=x+dx;
end
%Elements connectivity
nodel=zeros(nel,4); %matrix finite element
kel=0;kaux=0;
for i=1:ndx
  for j=1:ndy
    kaux=kaux+1;
    kel=kel+1;
    nodel(kel,1)=kaux+ndy+2;
    nodel(kel,2)=kaux+1;
    nodel(kel,3)=kaux;
    nodel(kel,4)=kaux+ndy+1;
  end
  kaux=kaux+1;
end
%problem dimensions
nglpn=2;%number of degrees of freedom per node
nds=nno*nglpn; %number of system displacements
nnel=4; %number of nodes per element
ndpel=nnel*nglpn;%number of displacements per element
%physical data of the elements
t0=0.1; %inital thickness
EM=200e9; %steel elasticity module
nu=0.3; %Poisson's ratio of steel
EL=EM/(1-nu*nu);
G=EM/2/(1+nu); %shear module
E=EL*[1 nu 0;nu 1 0;0 0 (1-nu)/2]; %Elasticity matrix
```

Then it will be sought to obtain the stresses present in each element. For that end, it is necessary to determine certain matrices and vectors. In this way, the loads, constraints, stiffness matrix and displacements are then defined.

```
%matrix: number of degrees of freedom per node
LN=zeros(nno,nglpn);
%boundary conditions
%cantilever beam
for i= k-ndy:k
   LN(i,:)=[-1 -1];
end
%matrix LN
ngl=0;
for i=1:nno
   for j=1:nglpn
      if LN(i,j)== 0
         ngl=ngl+1;
         LN(i,j)=ngl;
      end
   end
end
ngr=ngl;
for i=1:nno
   for j=1:nglpn
      if LN(i,j)<0
         ngr=ngr+1;
         LN(i,j)=ngr;
      end
   end
end
%matrix and vector initialization
K=zeros(nds,nds);
p=zeros(nds,1);
P=zeros(nds,1);
Tens=zeros(nel,3);
q=zeros(8,1);
% loading vector P
%vertical load
%V=-100e4;
% for i=1:ndy+1
%   P(LN(i,2))=V/(ndy+1);
% end
```

```
%stiffness matrices of the elements
nd=ones(1,4);
for iel=1:nel
  for j=1:nnel
    nd(j)=nodel(iel,j);
  end
  xa=gcoord(nd(1),1);xb=gcoord(nd(2),1);
  yb=gcoord(nd(2),2);yc=gcoord(nd(3),2);
  t = vt(iel,1);
%rectangle dimensions
  a=(xa-xb)/2;
  b=(yb-yc)/2;
%constants
  c1=EL*t*b/3/a;
  c2=c1/2;
  c3=EL*t*nu/4;
  c4=G*t*a/3/b;
  c5=c4/2;
  c6=G*t/4;
%
  kd(1,1)=c1;kd(1,2)=c3;kd(1,3)=-c1;kd(1,4)=c3;kd(1,5)=-c2;kd(1,6)=-c3;
kd(1,7)=c2;kd(1,8)=-c3;
  kd(2,2)=c1;kd(2,3)=-c3;kd(2,4)=c2;kd(2,5)=-c3;kd(2,6)=-c2;kd(2,7)=c3;
kd(2,8)=-c1;
  kd(3,3)=c1;kd(3,4)=-c3;kd(3,5)=c2;kd(3,6)=c3;kd(3,7)=-c2;kd(3,8)=c3;
  kd(4,4)=c1;kd(4,5)=-c3;kd(4,6)=-c1;kd(4,7)=c3;kd(4,8)=-c2;
  kd(5,5)=c1;kd(5,6)=c3;kd(5,7)=-c1;kd(5,8)=c3;
  kd(6,6)=c1;kd(6,7)=-c3;kd(6,8)=c2;
  kd(7,7)=c1;kd(7,8)=-c3;
  kd(8,8)=c1;
%
  ks(1,1)=c4;ks(1,2)=c6;ks(1,3)=c5;ks(1,4)=-c6;ks(1,5)=-c5;ks(1,6)=-c6;
ks(1,7)=-c4;ks(1,8)=c6;
  ks(2,2)=c4;ks(2,3)=c6;ks(2,4)=-c4;ks(2,5)=-c6;ks(2,6)=-c5;ks(2,7)=-c6;
ks(2,8)=c5;
  ks(3,3)=c4;ks(3,4)=-c6;ks(3,5)=-c4;ks(3,6)=-c6;ks(3,7)=-c5;ks(3,8)=c6;
  ks(4,4)=c4;ks(4,5)=c6;ks(4,6)=c5;ks(4,7)=c6;ks(4,8)=-c5;
  ks(5,5)=c4;ks(5,6)=c6;ks(5,7)=c5;ks(5,8)=-c6;
  ks(6,6)=c4;ks(6,7)=c6;ks(6,8)=-c4;
  ks(7,7)=c4;ks(7,8)=-c6;
  ks(8,8)=c4;
%
```

```
    k=kd+ks;
%symmetry
for i=2:8
    for j=1:i-1
        k(i,j)=k(j,i);
    end
end
%sum in the system stiffness matrix
  kl=0;
  d = ones(1,8);
  for n=1:nnel
    kl=kl+1;
    d(kl)=LN(nd(n),1);
    kl=kl+1;
    d(kl)=LN(nd(n),2);
  end
  for i=1:ndpel
    for j=1:ndpel
      K(d(i),d(j))=K(d(i),d(j))+k(i,j);
    end
  end
end
%System solution
%calculation of displacements
disp('displacements')
p(1:ngl)=K(1:ngl,1:ngl)\(P(1:ngl)-K(1:ngl,ngl+1:nds)*p(ngl+1:nds));
disp(p)
%calculation of support reactions
disp('Esforços Nodais inclusive reacoes de apoio')
P(ngl+1:nds)=K(ngl+1:nds,1:ngl)*p(1:ngl)+K(ngl+1:ngl+1,ngl+1:ngl+1)..
    *p(ngl+1:nds);
disp(P)
%Stress
disp('Stresses at the central point of the elements')
disp('sigma_x,   sigma_y,   tau_xy')
for iel=1:nel
  for j=1:nnel
    nd(j)=nodel(iel,j);
  end
  xa=gcoord(nd(1),1);xb=gcoord(nd(2),1);
  yb=gcoord(nd(2),2);yc=gcoord(nd(3),2);
% rectangle dimensions
```

```
  a=(xa-xb)/2;b=(yb-yc)/2;
% constants
  ca=1/4/a;cb=1/4/b;
% matrix B=L*N calculated in the center of the element x=y=0
  B=[ca 0 -ca 0 -ca 0 ca 0;0 cb 0 cb 0 -cb 0 -cb;cb ca cb -ca -cb -ca -cb ca];
%
  kl=0;
  for n=1:nnel
    kl=kl+1;
    d(kl)=LN(nd(n),1);
    kl=kl+1;
    d(kl)=LN(nd(n),2);
  end
  for i=1:ndpel
    q(i)=p(d(i));
  end
  tau=E*B*q;
  Tens(iel,:)=tau';
end
disp(Tens)
```

Subsequent to obtaining the stresses, in order to have an optimization analysis based on the maximum allowable stresses, the maximum and minimum principal stresses and the maximum shear stresses are obtained, based on the stresses present in each element with components in x and y as follows:

$$\sigma_{1,2} = \frac{\sigma_x + \sigma_y}{2} \pm \sqrt{\left(\frac{\sigma_x + \sigma_y}{2}\right)^2 + \tau_{yx}^2} \qquad (8.7)$$

$$\tau_{max} = \sqrt{\left(\frac{\sigma_x + \sigma_y}{2}\right)^2 + \tau_{yx}^2} \qquad (8.8)$$

```
%Normal Stress at each element [MPa]
%sigma_1,  sigma_2,  sigma_max
smax = zeros(3,3);
for i=1:nel
  smax(i,1)=(Tens(i,1)+Tens(i,2))/2+(((Tens(i,1)-Tens(i,2))/2)^2+Tens
(i,3)^2)^0.5;
  smax(i,2)=(Tens(i,1)+Tens(i,2))/2-(((Tens(i,1)-Tens(i,2))/2)^2+Tens
(i,3)^2)^0.5;
  if abs(smax(i,1))>abs(smax(i,2))
    smax(i,3)=smax(i,1);
  else
```

```
   smax(i,3)=smax(i,2);
  end
end
for i=1:nel
  smax(i,3)=abs(smax(i,3));
end
%disp(smax)
%vector maximum stress [MPa]
vmax = zeros(nel,1);
for i=1:nel
  vmax(i)= smax(i,3);
end
```

8.3 The KINITRO algorithm

Now there is an interesting basis that will allow us to program the optimization process of the structure and for that we will turn to a theoretical scheme of the whole process.

Initially a classic optimization problem can be defined as: Determine $\mathbf{x} \in \Re^n$ that minimizes the objective function $f(\mathbf{x})$ subject to

$$\text{Equality constraints:} \quad g_j(\mathbf{x}) = 0; \quad j = 1, l \tag{8.9}$$

$$\text{Inequality constraints:} \quad g_j(\mathbf{x}) \le 0; \quad j = l+1, \ m \tag{8.10}$$

The Lagrangian function of the problem is defined as

$$\Lambda(\mathbf{x}, \mathbf{u}) = f(\mathbf{x}) + \sum_{i=1}^{m} u_i g_i(\mathbf{x}) \tag{8.11}$$

where $\mathbf{u} \in \Re^m$ is the vector of the Lagrange multipliers.

The variables, individually, represent certain factors of a certain project (Brasil and Silva 2018). These values are independent of each other and constantly changed during the optimization process as requested by the problem-solving tool. In this case, the design variables are the thickness vector (v_t), represented previously by the vector $\mathbf{x}$, and the number of divisions in x and y directions:

$$v_{t\,(k,1)} = [vt_1\ vt_2 \ldots], \quad k = 1, \ldots, \text{nel} \tag{8.12}$$

Constraints are a set of limitations imposed on the system. They can be of inequality and equality, indicating maximum or minimum values that should not be exceeded. Here, the restrictions are nonlinear inequalities and refer to the allowable stress, $\bar{\sigma} = 225 \times 10^6 \text{N}$, of the material and minimum thickness, $v_{\text{tmin}} = 0.0001\text{m}$.

$$g_1 = \bar{\sigma}_{\max} \le \sigma_{\text{adm}} = \sigma_{\max} - \bar{\sigma} \tag{8.13}$$

$$g_2 = V_t \geq V_{\text{tmin}} = V_{\text{tmin}} - V_t \qquad (8.14)$$

The objective function $f(\mathbf{x})$ corresponds to a single value, connected with the whole project. It must optimize the project in order to maximize it, minimize it, or reach a desired value. The volume is the most important data of the project, since from it, the quantity of material to be used is known. It is understood that by optimizing the volume, that is, by minimizing it, the entire project is optimized. Therefore, the volume (*Vol*) will be the objective function:

$$\text{Area}_{(1,k)} = \begin{bmatrix} dx_1 * dy_1 \\ \cdots \end{bmatrix}, \quad k = 1, \ldots, \text{nel} \qquad (8.15)$$

$$\text{Vol} = \text{Area} * v_t \qquad (8.16)$$

To solve the problem, the KNITRO (Artelys 2021) algorithm will be used, which has an easy coding interface for MATLAB. The function returns, among other parameters, the optimal value of the design variables, the value of the function, Lagrange multipliers, and so on. The illustration of the arguments is shown in Figure 8.3. It is worth mentioning the importance of choosing a good algorithm, because without one "the chessboard irregularity" can occur (Bensoe and Sigmund 2003), due to the high dependence of the algorithm on all the parameters of the project. Therefore, a good choice allows us to be able to maintain good results, regardless of initial values, or any other parameter, thus providing solid and consistent results.

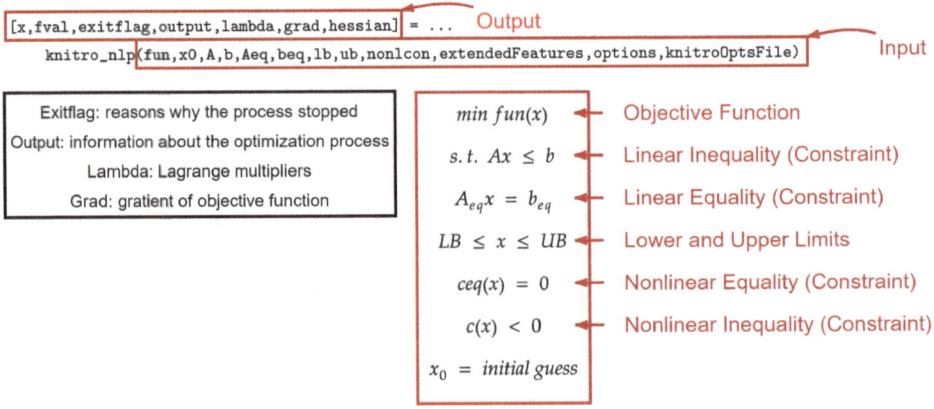

Figure 8.3: Arguments of the KNITRO function in MATLAB (Artelys, 2021).

Based on what was described, the initial formulation of the problem for optimization in MATLAB follows the given format:

```
vol = @(vt) Area*vt;
options = knitro_options('outlev',3);
[vt,fval,exitflag,output] = knitro_nlp(vol,x0,[],[],[],[],[],[],@nonlcon,
[],options,[]);
```

It is important to note that "*@nonlcon*" is performed in a function script that admits a set of initial values and returns two vectors $c(x)$ and $c_{eq}(x)$. In addition, "options" (Artelys, 2021) has a series of options capable of not only modifying how the optimization module operates but also how the information is made available. In this case we have an "outlev" capable of informing the amount of process information that will be available to the user, such as providing a quick summary of the process to all information, including all iterations.

Note that the KNITRO function has several input and output parameters. The optimization problem shown in Figure 8.3 and in the initial formulation is the same governed by Eqs. (8.9)–(8.11), written in a more detailed way, separating linear from non-linear restrictions, as well as the lower and upper limits of the function.

After the optimization process, as a way of presenting the results, a "GRID," mesh or chessboard will be made, using the "gray scale," expressing in mathematical and visual terms both the maximum stress and the thickness, in a similar way as shown in Figure 8.1, and programmed as follows:

```
%Generate the X and Y grid arrays using the MESHGRID function.
clf;
%Maximum Stress
x = (1:ndx+1);
y = (1:ndy+1);
[X,Y] = meshgrid(x,y);
%Thickness
x1 = (1:ndx+1);
y1 = (1:ndy+1);
[X1,Y1] = meshgrid(x1,y1);
%Generator of Matrix Z
Z = ones(ndy+1,ndx+1);
Z1 = ones(ndy+1,ndx+1);
kx=0;
for j=1:ndx
  for i=1:ndy
    kx=kx+1;
    Z(i,j)= vmax(kx);
    Z(ndy+1,j)= 0;
    Z(i,ndx+1)= 0;
    Z(ndy+1,ndx+1)=0;
```

```
    Z1(i,j)= vt(kx);
    Z1(ndy+1,j)= 0;
    Z1(i,ndx+1)= 0;
    Z1(ndy+1,ndx+1)=0;
  end
end
%Plots
%Maximum Stress
subplot(2,1,1)
s = pcolor(X,Y,Z);
title('Maximum Stress')
colormap(flipud(gray));
colorbar;
axis image;
%Thickness
subplot(2,1,2)
s1 = pcolor(X1,Y1,Z1);
title('Thickness')
colormap(flipud(gray));
colorbar;
axis image;
```

There are some relevant points to note before presenting the results obtained in the optimization. It is known that the finer the discretization of the mesh, the more accurate the results will be. However, the constant increase results in an almost exponential computational cost. For this reason, discretization was established in two and three times of the initial dimensions of the structure, which have an initial thickness of 0.1 m.

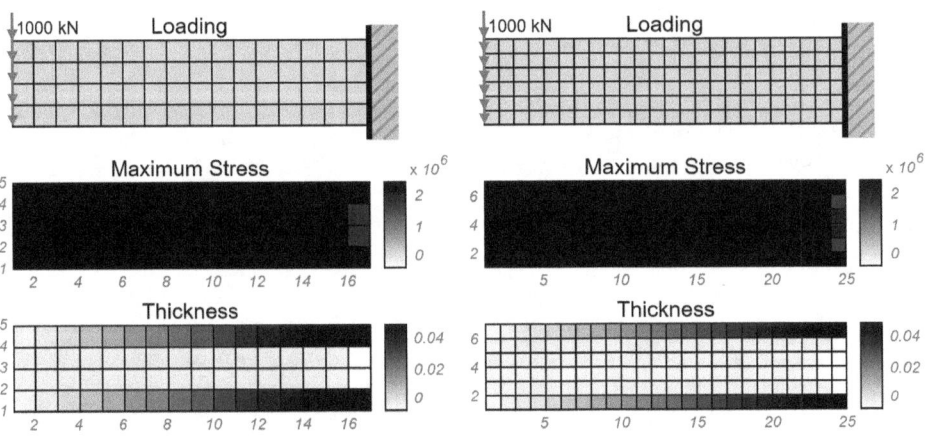

Figure 8.4: Cantilever plate with concentrated loading on the edge (×2 and ×3).

Table 8.1: Cantilever plate with concentrated loading on the edge.

Dimensions	8 × 2	
Initial volume	1.6 m³	
	×2	×3
Final volume	0.201368 m³	0.194917 m³
Reduction	87.41%	87.82%

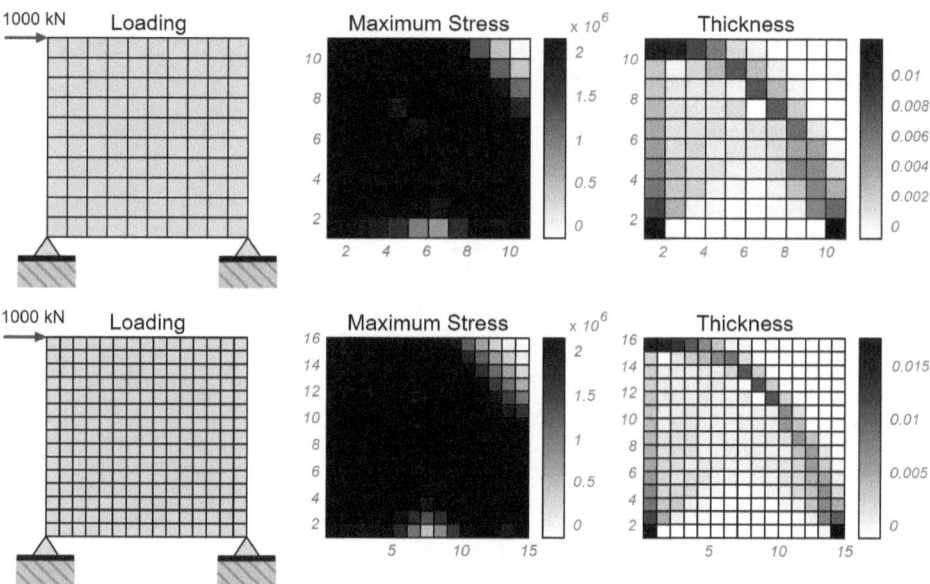

Figure 8.5: Simply supported plate with concentrated side-loading ×2 and ×3).

Table 8.2: Cantilever plate with concentrated loading on the edge.

Dimensions	5 × 5	
Initial volume	2.5 m³	
	×2	×3
Final volume	0.055845 m³	0.056045 m³
Reduction	97.77%	97.76%

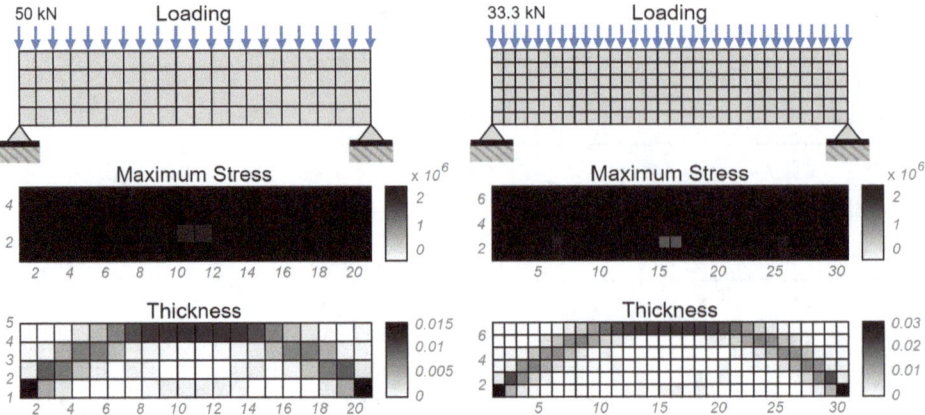

Figure 8.6: Simply supported beam with uniformly distributed loading (×2 and ×3).

Table 8.3: Double-based beam with uniformly distributed loading.

Dimensions	10 × 2	
Initial volume	2 m³	
	×2	×3
Final volume	0.072808 m³	0.108060 m³
Reduction	96.36%	94.60%

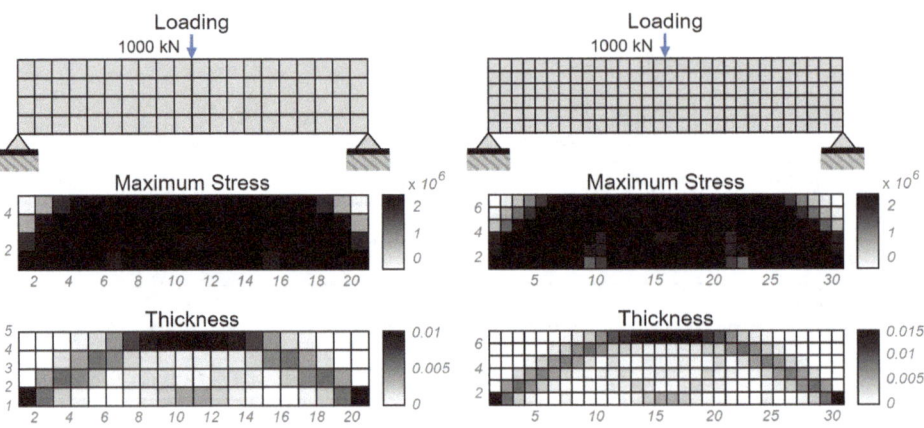

Figure 8.7: Simply supported beam with concentrated centered loading (×2 and ×3).

Table 8.4: Simply supported beam with concentrated centered loading.

Dimensions		10 × 2	
Initial volume		2 m³	
	×2		×3
Final volume	0.061673 m³		0.061700 m³
Reduction	96.92%		96.91%

8.4 Reliability

After programming the entire optimization process and obtaining the results, it is possible to add a highly relevant aspect, which is the structural reliability.

Reliability is understood as the ability of an equipment or human being to perform their expected functions properly under specific conditions during a given period of time, in the absence of breaks or failures. It is an area of study that aims to evaluate and optimize the reliability of systems through techniques derived from probability and statistics theories. In history, the concept of reliability acquired technological significance after the end of the First World War, when it was used for comparative studies carried out on airplanes with one, two, or four engines, in order to measure the number of accidents per flight hour. It was during the Second World War that, as a result of a failure of German V-1 missiles, mathematician Robert Lusser proposed a probability law for a serial component product. Only in 1963, a first association brought together engineers from the reliability sector and the first periodical for the dissemination of works in the area appeared in the United States.

A general problem is built around the idea of "discrete events," being developed to help follow a model over time, since it often involves a complex logical structure of its elements (Jin 1993) necessary to determine the relevant amounts that are of interest. The simulation based on this structure is often called simulation of discrete events. Therefore, the study of the Monte Carlo method is a great alternative for such simulations and requires the understanding of different areas of knowledge: probability, to describe processes and random experiments; statistics, to analyze the data; computer science, for efficient implementation of algorithms; and mathematical programming, to formulate and solve optimization problems.

The computational method uses random numbers and statistics to solve problems, since currently several numerical problems in finance, engineering, and statistics are solved with the Monte Carlo method. The interest in this study is to apply the technique in order to make the budget of a structural project feasible in terms of mass.

A limit mass, m_{el}, will be obtained based on the available budget. In the case of structural reliability analysis (Nowak and Collins 2012), this means, in the simplest approach, sampling each random variable to provide a sample value. With changes in the allowable stress and using the optimization process described in this chapter, it is possible to obtain the mass of the structure, m_e. Therefore, using Eq. (8.17), we can interpret the situation.

$$M = m_{el} - m_e \tag{8.17}$$

Equation (8.17) is then verified using the sample value set. If the function is violated (i.e., M < 0), the structure or structural element has "failed (Melchers and Beck 2018). This procedure will be run n times in order to obtain a considerable set of values referring to M, so being possible to obtain its mean ($\bar{M}$) and standard deviation ($\hat{M}$). With this whole set of data, it is possible to calculate the reliability index (β), a factor that measures the distance between the origin and the average value present in a normal distribution as follows:

$$\beta = \frac{\bar{M}}{\hat{M}} \tag{8.18}$$

Based on the reliability index, the probability of failure can be calculated as follows:

$$P = \Phi(-\beta) \tag{8.19}$$

where Φ is the standard normal cumulative function.

$$\Phi(y) = \frac{1}{\sqrt{2\pi}} \int_{-\infty}^{y} e^{-\frac{z^2}{2}} dz \tag{8.20}$$

However, there are often hardware limitations, so it is necessary to use other methods besides Monte Carlo, due to the fact that it has a high computational cost. Consequently, the classic method for calculating the probability of failure is also used:

$$P_f = \frac{\text{Non favorable cases}}{\text{Total cases}} \tag{8.21}$$

A hundred iterations were performed for the processes and the limit mass was based on the final mass obtained in the optimization process. Using the final value of the volume and the density of the material, it's possible to find the limit mass, being initially increased by 10% of the base value, as follows:

$$m_{el} = (\text{Vol} \times 7,800) \times (1 + 10\%) \tag{8.22}$$

```
%limit mass
porcent = input('Percentage for the limit mass ? ');
mel = (fval*7800)*(1+(porcent/100));
Minter = input('Amount of iterations ? ');
```

```
me = ones(Minter,1);
tic;
for h=1:Minter
  csadm = norminv(rand(),260e6,26e6);
  reliability; %script to run the optimization process
  me(h,1) = (fval*7800); %[kg]
end
time2=toc;
%M
M = mel - me;
%average
M_mean = mean(M);
%standard deviation
M_std = std(M);
%beta
beta = (M_mean)/(M_std);
%standard normal cumulative function
fun2 = @(z) exp(-(((z).^2)/2));
int_fun2 = integral(fun2,-Inf,-beta);
P_phi = (1/(sqrt(2*pi)))*int_fun2;
T4 = table(beta,P_phi,time2,...

'VariableNames',{'Beta','FailureProbability','Runtime',},...
'RowNames',{'Results'});
disp(T4)
%Classic Method - Reliability
cont = 0;
for i = 1:Minter
  if(M(i)<0)
            cont = cont +1;
  end
end
falha2 = cont/Minter;
probabilidadefalha2 = falha2*100;
T5 = table(probabilidadefalha2,...
  'VariableNames',{'ProbabilidadeFalha2'},...
  'RowNames',{'Resultado (Confiabilidade2)'});
disp (T5)
```

Table 8.5: Cantilever plate with concentrated loading on the edge.

	Monte Carlo method	
	×2	×3
Beta	2.6553	2.72879
Failure probability	0.39614%	0.31871%
	Classic method	
	×2	×3
Failure probability	1%	2%
Runtime	0.26546 h	2.40272 h

Table 8.6: Simply supported plate with concentrated side-loading.

	Monte Carlo method	
	×2	×3
Beta	2.5092	2.9796
Failure probability	0.60507%	0.14430%
	Classic method	
	×2	×3
Failure probability	2%	0%
Runtime	2.0745 h	27.9 h

Table 8.7: Simply supported beam with uniformly distributed loading.

	Monte Carlo method	
	×2	×3
Beta	2.844	2.674
Failure probability	0.22273%	0.37477%
	Classic method	
	×2	×3
Failure probability	0%	1%
Runtime	0.57631 h	12.235 h

Table 8.8: Simply supported beam with concentrated centered loading.

	Monte Carlo method	
	×2	×3
Beta	2.5157	2.6779
Failure probability	0.59400%	0.37040%
	Classic method	
	×2	×3
Failure probability	1%	1%
Runtime	0.48483 h	13.885 h

Chapter 9
Using Excel Solver in optimization problems

Solver is a supplement to Microsoft Excel (or simply Solver) and can be used to solve optimization problems. The main types of problems solved are:
- linear programming (LP) with Simplex
- nonlinear programming with the generalized reduced gradient method (GRG)
- evolutionary

In case of nonlinear continuous problems, Solver uses a version of the GRG. The GRG method is based on quadratic programming. The idea is to find a search direction such that the active constraints remain ε-active for small movements in the space of the design variables and use Newton Raphson's method to return to the bounds of the constraints when they are not ε-active. Excel's GRG offers both the conjugate and Newton gradients method to determine the search direction. The Simplex method was explained in Chapter 4. The evolutionary methods were approached in Chapter 7. In this chapter practical examples of linear and nonlinear optimization problems will be described.

9.1 Installing the Excel Solver

To have the solver available in your Microsoft Excel just follow the steps described in the sequence. Load the Solver Add-in in Excel:
1. In Excel 2010 and later, go to File > Options. ...
2. Click Add-Ins, and then in the Manage box, select Excel Add-ins.
3. Click Go.
4. In the Add-Ins available box, select the Solver Add-in check box, and then click OK. ...
5. After you load the Solver Add-in, the Solver command is available in the Analysis group on the Data tab.

9.2 The Solver's window

The Solver communication window and commands are shown in Figure 9.1. Description of the main communication windows is as follows:
1. In the window of the objective function cell, the Excel spreadsheet cell should be chosen where the equation of the objective function is written.
2. In type of optimization option should be chosen if the goal is to minimize, maximize or a given value for the objective function.

https://doi.org/10.1515/9783110625622-009

3. In the "By Changing Variable Cells" window, the cells where the design variables are located, that is, the input cells of the design variables, should be indicated; in these cells there should be no formulas, but input numbers.
4. In the "Subject to Constraints" window, the equations of the constraints should be described; you have the option here to add, change, or remove a particular constraint; as a suggestion, avoid using mathematical expressions in this window, write the expressions in the Excel spreadsheet, and here use only expressions like a certain cell "=0" or "≤0" or "≥0."
5. In optimization method, one of the methods, LP, GRG, or Discrete Optimization, should be chosen; In "options" there are several options both for numerical calculation and for some methods to be used.
6. After defining the entire formulation click on Solve to trigger the Solver and have the solution.

For more information, please refer to the Microsoft Excel help.

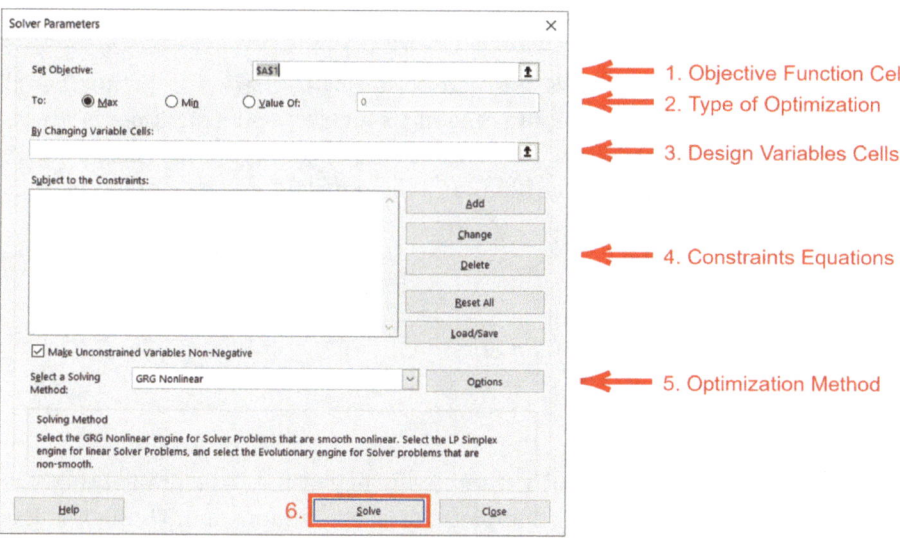

Figure 9.1: Communication window and Excel Solver commands.

9.3 Example 1: using Solver to calculate the eigenvalue of a structural dynamic problem

Given a structure, like that one shown in Figure 9.2, from its geometric and mechanical properties it is possible to obtain its natural vibration frequencies.

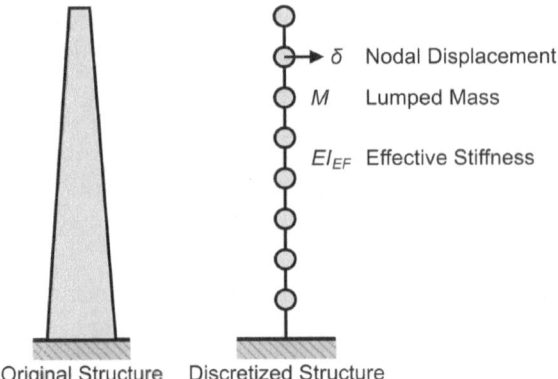

Original Structure   Discretized Structure

Figure 9.2: Discretized structure of a telecommunication tower.

The natural frequencies of vibration are determined with the solution of the characteristic equation:

$$\det(\mathbf{K} + \lambda\mathbf{M}) = 0 \tag{9.1}$$

In this equation, $\mathbf{K}$ is the stiffness matrix, of order $n x n$, where n is the number of degrees of freedom of the system, $\mathbf{M}$ is the mass matrix, also $n x n$, and λ is the eigenvalue. Note that Eq. (9.1) is a polynomial in λ of order n. Therefore, it has n roots. The roots λ_i are called eigenvalues and the natural frequency of vibration is given by

$$\omega_i = \sqrt{\lambda_i} \tag{9.2}$$

where ω_i is called circular frequency and

$$f_i = \frac{\omega_i}{2\pi} \tag{9.3}$$

cyclic frequency. The lowest (first) natural frequency of the structure is called the fundamental frequency, which is an important piece of information. Through it, it is possible to verify the need or not to carry out a dynamic analysis, in the case of structures subjected to wind loading, as well as the "risk" of an eventual resonance in case of a machine foundation project.

Next, it is shown how to compute the lowest eigenvalue and the fundamental frequency of the structure of a steel wind energy tower. Table 9.1 shows a brief description of the geometry characteristics of this tower, where H is the height above ground of the structure, Diam. tip and Diam. base are the diameters of the cross-section at the top and button respectively, while Thic. tip and Thic. base are the thicknesses at the top and button, respectively.

Table 9.1: Wind energy tower geometry.

H:	60.0	m			
Diam. tip:	200.00	cm	Diam. base:	708.04	cm
Thic. tip:	0.93	cm	Thic. base:	1.03	cm

In addition to the geometry of the structure that must be used to determine the stiffness matrix, the density of 7,850 kg/m^3 and a mass concentrated on top of 104,000 kg must be considered. For the calculation, the structure was discretized into 40 elements of equal length. Each node has two degrees of freedom: one is the horizontal displacement and the other is the rotation around a horizontal axis perpendicular to the plane of Figure 9.2. With this, it is possible to determine the stiffness matrix **K** and mass **M** of the discretized structure.

Table 9.2 shows, among others, the objective function, and the design variable for the initial design. Note that lamb is the eigenvalue and design variable, f is calculated according to Eqs. (9.2) and (9.3), Det is the determinant shown in Eq. (9.1). Note that the value of Det is different from zero and at the optimum point it should be equal to zero, or a small number, for the adopted precision. The fundamental frequency calculated with a program based on the Finite Element Method (FEM) is f-fem, Error approx is the percentage error between the value of f and f-fem, and scale is a scale factor used to prevent the determinant value from being too large or too small.

Table 9.2: Initial design.

Computation of the first eigenvalue			
lamb =	40.000	(rad/s)^2	(initial eigenvalue)
f =	1.01	Hz	(initial fundamental frequency)
Det =	−1.40E + 02		(residue of the characteristic equation)
scale =	8		(factor scale)
f-fem =	0.62	Hz	(frequency computed using FEM)
Error aprox =	61.2363%		(error between Excel Solver and FEM)

Table 9.3 shows the final design obtained. Note that the determinant of the characteristic equation is a small number, close to zero, and the approximation error of the fundamental frequency value is equal to 10^{-7}, a very satisfactory result (excellent).

Table 9.3: Final design obtained by Excel Solver.

Computation of the first eigenvalue			
lamb =	15.386	$(rad/s)^2$	(final eigenvalue)
f =	0.62	Hz	(final fundamental frequency)
Det =	1.78E-05		(residue of the characteristic equation)
scale =	8		(factor scale)
f-fem =	0.62	Hz	(frequency computed using FEM)
Error aprox =	−0.00001%		(error between Excel Solver and FEM)

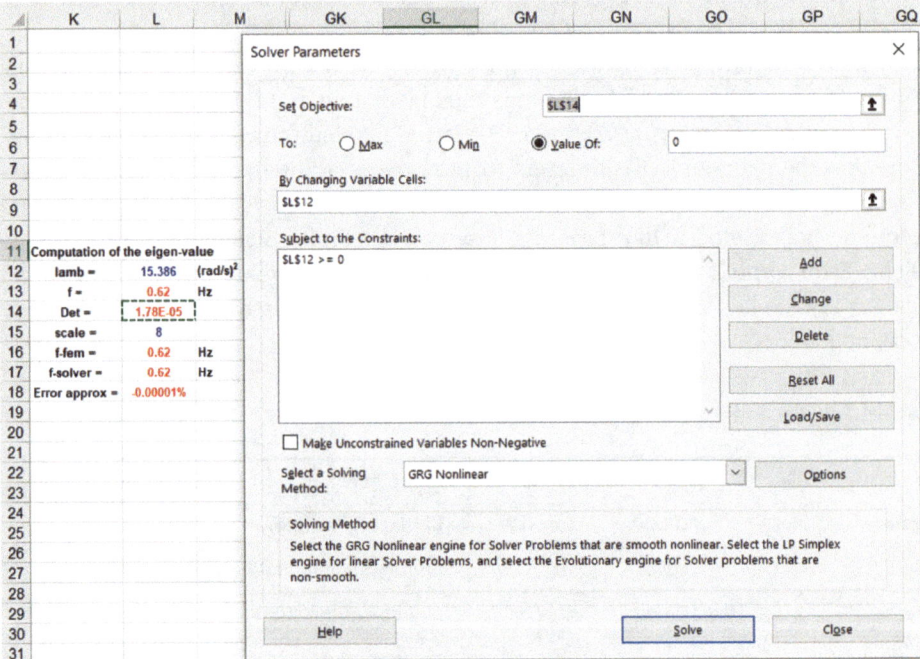

Figure 9.3: Solver communication window.

The Solver communication window for the problem is shown in Figure 9.3. Observe in this figure that cell L14 (Det) is the objective function, which must be null, cell L12 is the design variable, and the imposed constraint is that the design variable must be positive. It was also decided to use the GRG to solve the optimization problem. The optimum design obtained is $\mathbf{x}^\star = [0.62]$ and $f(\mathbf{x}^\star) = 1.78 \times 10^{-5} \cong 0$.

9.4 Solving Example 3.1 (Example E1 of Chapter 1): linear programming

The design variables are:

$$x_1 = \text{number of toys A}, \quad x_2 = \text{number of toys B}$$

Objective function (profit), to be maximized: $F(x) = 400x_1 + 600x_2$

Inequality constraints:

$$x_1 + x_2 \le 16 \quad \Rightarrow \quad g_1(x) = x_1 + x_2 - 16 \le 0 \qquad \text{(shipping)}$$

$$x_1/28 + x_2/14 \le 1 \quad \Rightarrow \quad g_2(x) = x_1/28 + x_2/14 - 1 \le 0 \quad \text{(production)}$$

$$x_1/14 + x_2/24 \le 1 \quad \Rightarrow \quad g_3(x) = x_1/14 + x_2/24 - 1 \le 0 \quad \text{(sales)}$$

Writing these expression in Excel, one can see the formulas in Figure 9.4.

B4		f_x =400*B1+600*B2		
	A	B	C	D
1	x1 =	0	Design variable 1	
2	x2 =	0	Design variable 2	
3				
4	F(x) =	=400*B1+600*B2	Profit	Objective Function
5				
6	g1 =	=+B1+B2-16	Constraint 1	
7	g2 =	=B1/28+B2/14-1	Constraint 2	
8	g3 =	=B1/14+B2/24-1	Constraint 3	

Figure 9.4: Expressions of Example 3.1.

Note that in cells B1 and B2 are the input of the design variables. These cells do not depend of any other one. In cell B4 the objective function is written, and finally, in cells B6 to B8 the expressions of the constraints are shown. As stated before, the ideal is to write the constraints in the Excel spreadsheet and in the Excel Solver window just use "=0" or "≤0" or "≥0."

B4		f_x =400*B1+600*B2				
	A	B	C	D	E	F
1	x1 =	0	Design variable 1			
2	x2 =	0	Design variable 2			
3						
4	F(x) =	0	Profit	Objective Function		
5						
6	g1 =	-16	Constraint 1			
7	g2 =	-1	Constraint 2			
8	g3 =	-1	Constraint 3			

Figure 9.5: Numerical values of Example 3.1 for the initial design.

In Figure 9.5 are shown the initial values of the design variables and the objective and constraint functions.

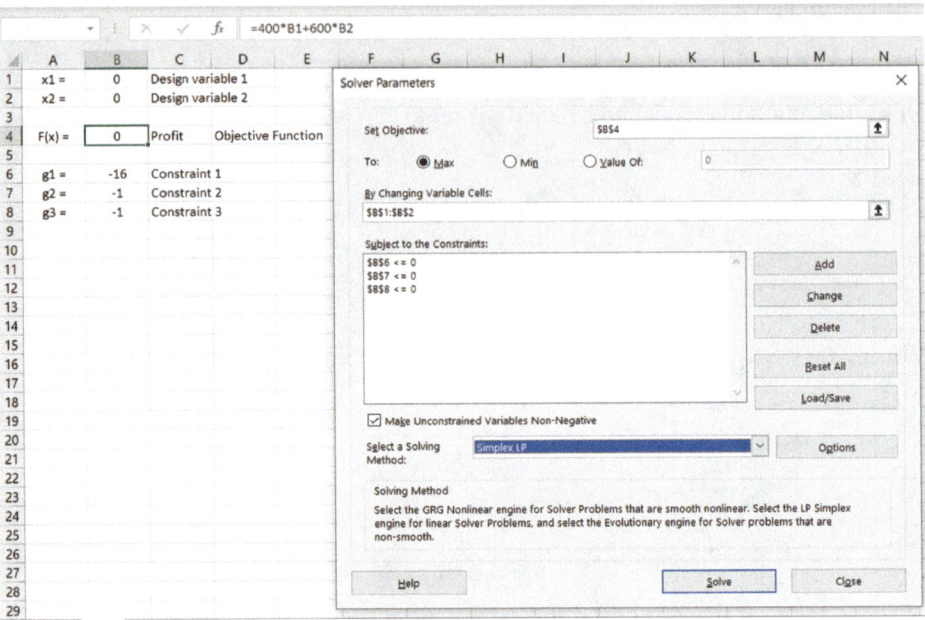

Figure 9.6: Excel Solver window, using Simplex LP, of Example 3.1.

In Figure 9.6 one can see the Excel Solver window. Note that the constraints are all written as "≤0." The Simplex LP was chosen to solve the problem and the obtained results are shown in Figure 9.7.

	A	B	C	D	E
1	x1 =	4	Design variable 1		
2	x2 =	12	Design variable 2		
3					
4	F(x) =	8800	Profit	Objective Function	
5					
6	g1 =	0	Constraint 1		
7	g2 =	0	Constraint 2		
8	g3 =	-0.21429	Constraint 3		

Figure 9.7: Numerical values of Example 3.1 for the final design.

Analyzing the data obtained in Figure 9.7 and those shown in Section 3.1 one could conclude that in both methods the final design is the same. The optimum design obtained is **x*** = [4 12] and $f(\mathbf{x}^*)$ = 8,800.

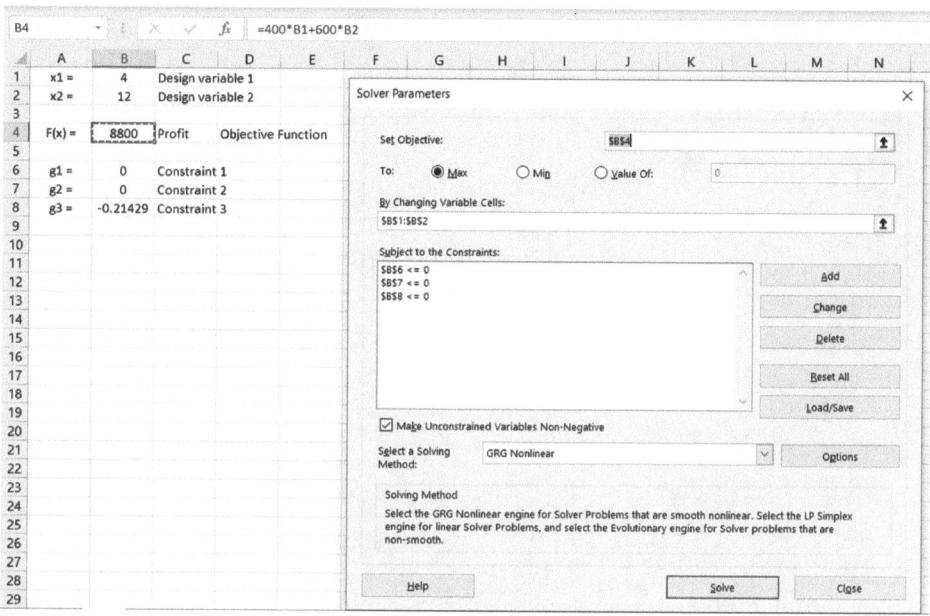

Figure 9.8: Excel Solver window, using GRG Nonlinear, of Example 3.1.

The problem can also be solved using GRG Nonlinear and the final design is the same as that shown in Figure 9.7. Observe that in both Simplex LP and GRG Nonlinear the box "Make unconstrained variables Non-Negative" is clicked. It is shown in Figure 9.8 the Excel Solver Window.

9.5 Solving Example 3.2: column under axial load

The column is shown in Figure 9.9. The length is $L = 5$ m. It is clamped in its base and free at the upper end. The section is tubular of average radius R and wall thickness t.

Other data are force $P = 10$ MN, elasticity modulus $E = 207$ GPa, density $\rho = 7833$ kg/m³, allowable stress $\sigma_a = 248$ MPa.

Design variables: R (x_1) and t (x_2)

Objective function: $f(R, t) = 2\rho L\pi Rt$, mass in kg

Inequality constraints:

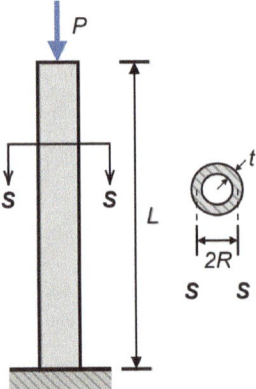

Figure 9.9: Structure of Example 3.2.

$$g_1(R, t) = \frac{P}{2\pi Rt} - \sigma_a \leq 0 \quad \text{(allowable stress)}$$

$$g_2(R, t) = P - \frac{\pi^3 E R^3 t}{4L^2} \leq 0 \quad \text{(buckling load)}$$

$$g_3(R, t) = -R \leq 0 \quad \text{(radius is positive)}$$

$$g_4(R, t) = -t \leq 0 \quad \text{(thickness is positive)}$$

As discussed in Section 9.3, this problem presents infinite solutions. The Excel Solver will be used to find one optimum point.

Writing these expression in Excel, one can see the formulas in Figure 9.10.

	A	B	C	D
1	x1 =	0.1666159452805	m	Average Radius
2	x2 =	0.0385169439351371	m	Thickness
3	ro =	7833	kg/m3	Density
4	L =	5	m	Length
5	E =	207	GPa	Elasticity Modulus
6	sa =	248	MPa	Allowable Stress
7	P =	10	MN	Axial Load
8				
9	F(x) =	=2*PI()*B1*B2*B4*B3	kg	Structure Mass
10				
11	g1 =	=B7/(2*PI()*B1*B2)-B6		Constraint 1
12	g2 =	=B7-PI()^3*B1^3*B2*B5*1000/(4*B4^2)		Constraint 2

Figure 9.10: Expressions of Example 3.2.

Cells B1 and B2 are the input of the design variables. In cell B9 the objective function is written, and finally, in cells B11 to B12 the expressions of the constraints are shown. Observe that constraints 3 and 4 do not need to be written in Excel spreadsheet; they can be imposed directly in the Solver window.

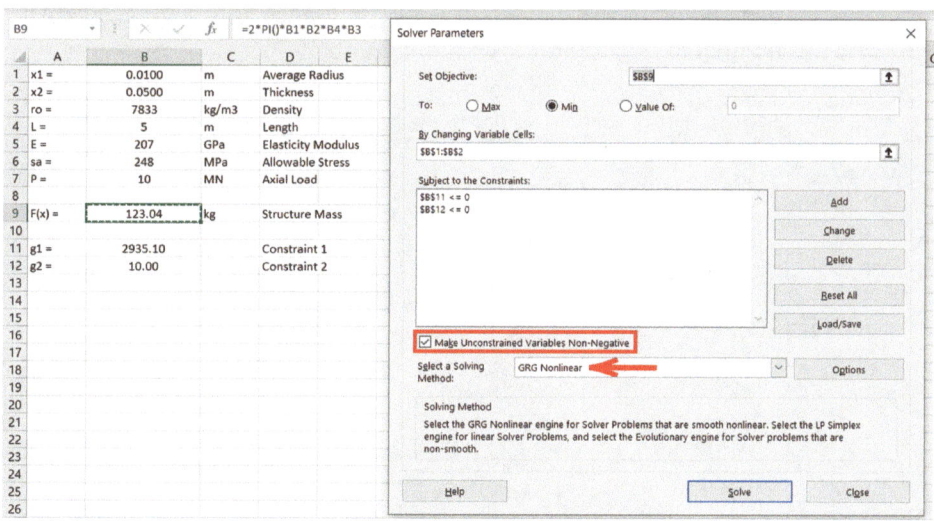

Figure 9.11: Numerical values of Example 3.2 for the initial design.

In Figure 9.11, the initial values of the design variables and the objective and constraint functions are shown.

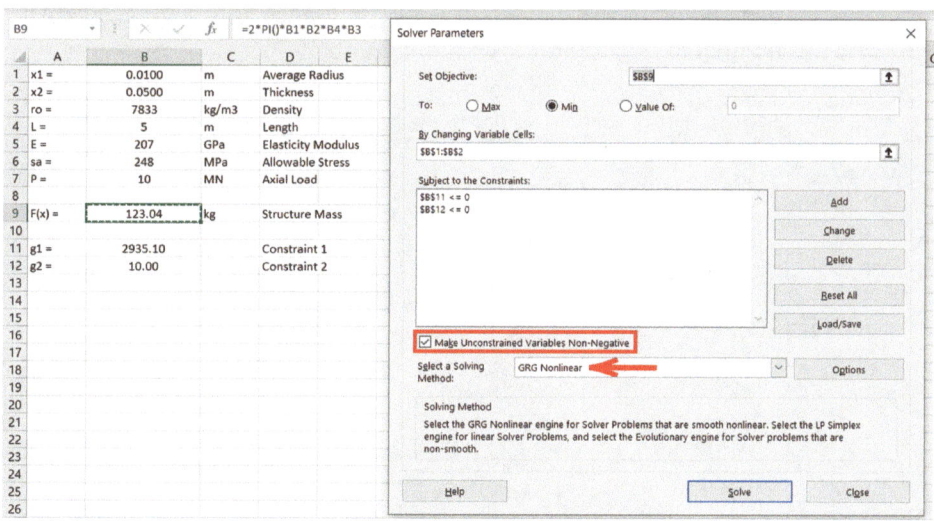

Figure 9.12: Excel Solver window, using GRG Nonlinear, of Example 3.2.

In Figure 9.12 one can see the Excel Solver window. Note that the constraints 3 and 4 can be replaced by just clicking "Make Unconstrained Variables Non-Negative" as shown in Figure 9.12. Other question is that this problem is nonlinear, so we must make the option of GRG Nonlinear.

The final design obtained, in Figure 9.13, is $\mathbf{x}^\star = [0.1603\ 0.0400]$ and $f(\mathbf{x}^\star) = 1579.23$. Observe that analyzing the data obtained in Figure 9.13 and those shown in

| B9 | ▼ | : | × | ✓ | fx | =2*PI()*B1*B2*B4*B3 |

	A	B	C	D	E
1	x1 =	0.1603	m	Average Radius	
2	x2 =	0.0400	m	Thickness	
3	ro =	7833	kg/m3	Density	
4	L =	5	m	Length	
5	E =	207	GPa	Elasticity Modulus	
6	sa =	248	MPa	Allowable Stress	
7	P =	10	MN	Axial Load	
8					
9	F(x) =	1579.23	kg	Structure Mass	
10					
11	g1 =	0.00		Constraint 1	
12	g2 =	-0.58		Constraint 2	

Figure 9.13: Numerical values of Example 3.2 for the final design.

Section 3.2, one can conclude that in both methods the final value of the objective function is the same, but the design is different. As stated, this problem presents infinite solutions. To find the same design we could introduce another constraint, imposing, for example, that $x_2 = 0.0405$. In Figure 9.14 one can see the Solver window with this additional constraint.

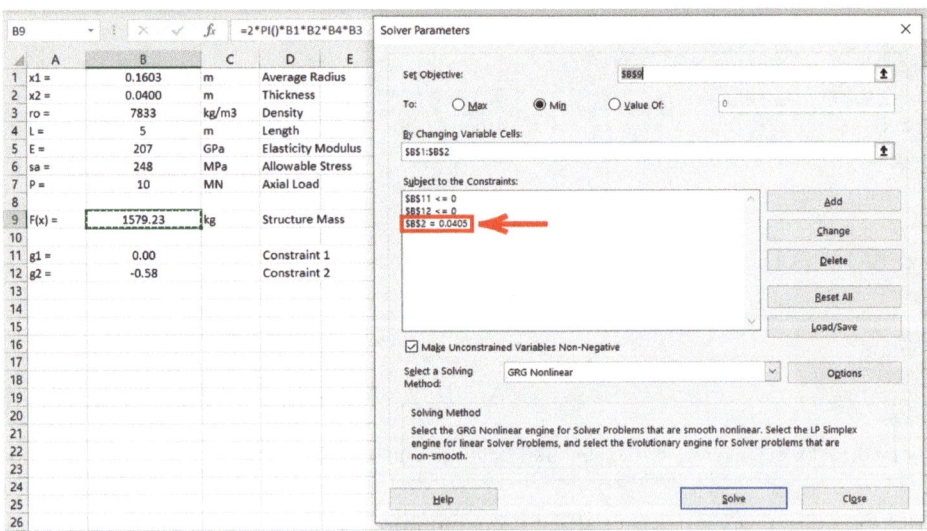

Figure 9.14: Excel Solver window, imposing $x_2 = 0.0405$ (Example 3.2).

The final design obtained, with this additional constraint, is shown in Figure 9.15. Observe that the value of the objective function is the same of that described in

Figure 9.13, but a different design. In this new solution the design vector is **x*** = [0.1575 0.0405] and $f(\mathbf{x}^*)$ = 1579.23, the same as that one presented in Example 3.2.

G9				f_x	
	A	B	C	D	E
1	x1 =	0.1575	m	Average Radius	
2	x2 =	0.0405	m	Thickness	
3	ro =	7833	kg/m3	Density	
4	L =	5	m	Length	
5	E =	207	GPa	Elasticity Modulus	
6	sa =	248	MPa	Allowable Stress	
7	P =	10	MN	Axial Load	
8					
9	F(x) =	1579.23	kg	Structure Mass	
10					
11	g1 =	0.00		Constraint 1	
12	g2 =	-0.34		Constraint 2	

Figure 9.15: Numerical values of the final design, imposing x_2 = 0.0405 (Example 3.2).

9.6 Solving Example 3.3: flexed beam

The example is shown in Figure 9.16, the beam in question. The problem is to minimize the beam mass. The data are:
- Length L = 10 m;
- Concentrated load P = 20 tf;
- Density ρ = 2500 kg/m³;
- Allowable stress σ_a = 20 MPa;
- Design variables $\mathbf{x}$ = [b_w h].

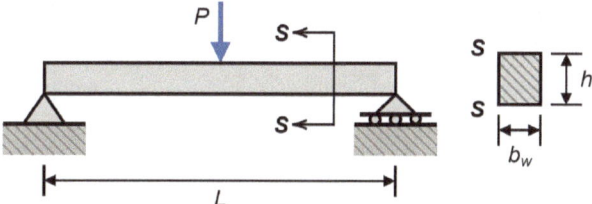

Figure 9.16: Simply supported beam, concentrated load at mid-span – Example 3.3.

Objective function: $f(b_w, h) = \rho L b_w h$, mass in kg
    Inequality constraints:

$$g_1(b_w, h) = \frac{3}{2}\frac{PL}{b_w h^2} - \sigma_a \leq 0 \quad \text{(allowable stress)}$$

$$g_2(b_w, h) = b_w - h \leq 0 \qquad \text{(height larger than width)}$$

$$g_3(b_w, h) = -b_w + 0.2 \leq 0 \quad \text{(minimum width)}$$

$$g_4(b_w, h) = -h + 0.2 \leq 0 \qquad \text{(minimum height)}$$

The expressions in the Excel spreadsheet can be seen in Figure 9.17. Cells B1 and B2 are the design variables. In cell B8 the objective function is written, and finally, in cells B10 to B13 the expressions of the constraints are shown.

B8		f_x	=B1*B2*B4*B3		
	A	B		C	D
1 x1 =		0.2		m	Width
2 x2 =		0.866025403784434		m	Height
3 ro =		2500		kg/m3	Density
4 L =		10		m	Length
5 sa =		20		MPa	Allowable Stress
6 P =		20		tf	Point Load
7					
8 F(x) =		=B1*B2*B4*B3		kg	Beam Mass
9					
10 g1 =		=3/2*B6*B4/(B1*B2^2)*10000-B5*1000000			Constraint 1
11 g2 =		=B1-B2			Constraint 2
12 g3 =		=-B1+0.2			Constraint 3
13 g4 =		=-B2+0.2			Constraint 4

Figure 9.17: Expressions of Example 3.3.

In Figure 9.18 the initial values of the design variables and the objective and constraints functions are shown.

B8		f_x	=B1*B2*B4*B3		
	A	B	C	D	E
1 x1 =		0.100	m	Width	
2 x2 =		0.100	m	Height	
3 ro =		2500	kg/m3	Density	
4 L =		10	m	Length	
5 sa =		20	MPa	Allowable Stress	
6 P =		20	tf	Point Load	
7					
8 F(x) =		250	kg	Beam Mass	
9					
10 g1 =		2980000000		Constraint 1	
11 g2 =		0.000		Constraint 2	
12 g3 =		0.100		Constraint 3	
13 g4 =		0.100		Constraint 4	

Figure 9.18: Numerical values of Example 3.3 for the initial design.

The window of the Excel Solver can be seen in Figure 9.19.

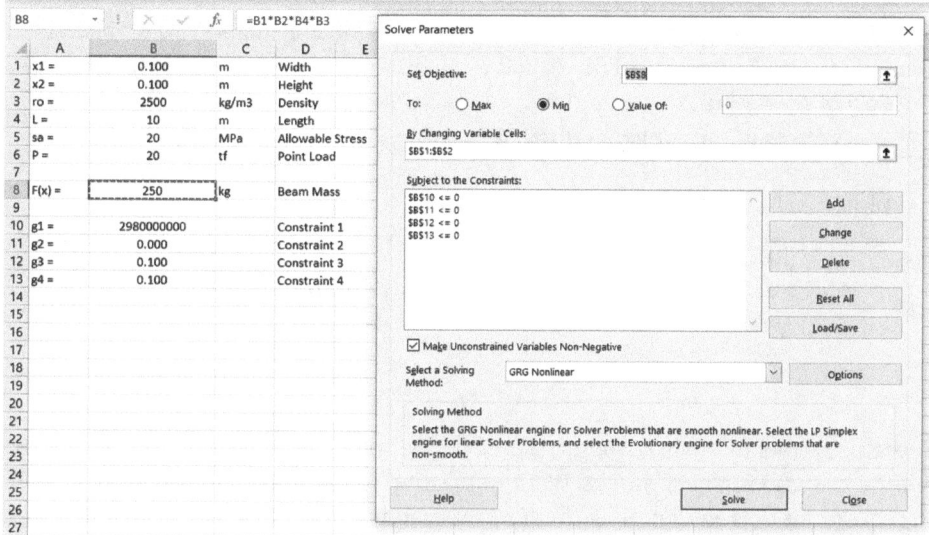

Figure 9.19: Excel Solver window of Example 3.3.

The final design obtained is shown in Figure 9.20, $\mathbf{x}^\star = [0.200\ 0.866]$ and $f(\mathbf{x}^\star) = 4{,}330$. Observe that the solution found here is the same of that shown in Section 3.3.

B8 f_x =B1*B2*B4*B3

	A	B	C	D	E
1	x1 =	0.200	m	Width	
2	x2 =	0.866	m	Height	
3	ro =	2500	kg/m3	Density	
4	L =	10	m	Length	
5	sa =	20	MPa	Allowable Stress	
6	P =	20	tf	Point Load	
7					
8	F(x) =	4330	kg	Beam Mass	
9					
10	g1 =	0.000		Constraint 1	
11	g2 =	-0.666		Constraint 2	
12	g3 =	0.000		Constraint 3	
13	g4 =	-0.666		Constraint 4	

Figure 9.20: Numerical values of Example 3.3 for the final design.

9.7 Solving Example 3.4: two electric generators

This example is about the design of two electric generators that are interconnected to provide a load of at least 60 units of power to a certain consumer. The problem is to compute the minimum cost of the P_1 and P_2 powers that each generator must provide.

Cost of power unit for generator 1: $C_1 = 1 - P_1 + P_1^2$

Cost of power unit for generator 2: $C_2 = 1 + 0.6P_2 + P_2^2$

Design variables: $x_1 = P_1$ e $x_2 = P_2$

Objective function: $f(x) = C_1 + C_2 = 2 - x_1 + x_1^2 + 0.6x_2 + x_2^2$

Subject to:

$$g_1 = -x_1 - x_2 + 60 \leq 0$$

$$g_2 = -x_1 \leq 0$$

$$g_3 = -x_2 \leq 0$$

The expressions in the Excel spreadsheet can be seen in Figure 9.21. Cells B1 and B2 are the design variables. In cell B5 the objective function is written and in cell B7 the constraint g_1 is shown. Note that for constraints 2 and 3 it is just needed to check the box "Make Unconstrained Variables Non-Negative."

B5		▾	⋮	✕	✓	f_x	=2-B1+B1^2+0.6*B2+B2^2	

	A	B	C	D
1	x1 =	10		Power Generator 1
2	x2 =	10		Power Generator 2
3	n =	60		Units
4				
5	F(x) =	=2-B1+B1^2+0.6*B2+B2^2		Cost Function
6				
7	g1 =	=-B1-B2+60		Constraint 1

Figure 9.21: Expressions of Example 3.4.

In Figure 9.22 the initial values of the design variables and the objective and constraint functions are shown.

B5		▾	⋮	✕	✓	f_x	=2-B1+B1^2+0.6*B2+B2^2		

	A	B	C	D	E	F	G
1	x1 =	10.0		Power Generator 1			
2	x2 =	10.0		Power Generator 2			
3	n =	60		Units			
4							
5	F(x) =	198		Cost Function			
6							
7	g1 =	40		Constraint 1			

Figure 9.22: Numerical values of Example 3.4 for the initial design.

The window of the Excel Solver can be seen in Figure 9.23.

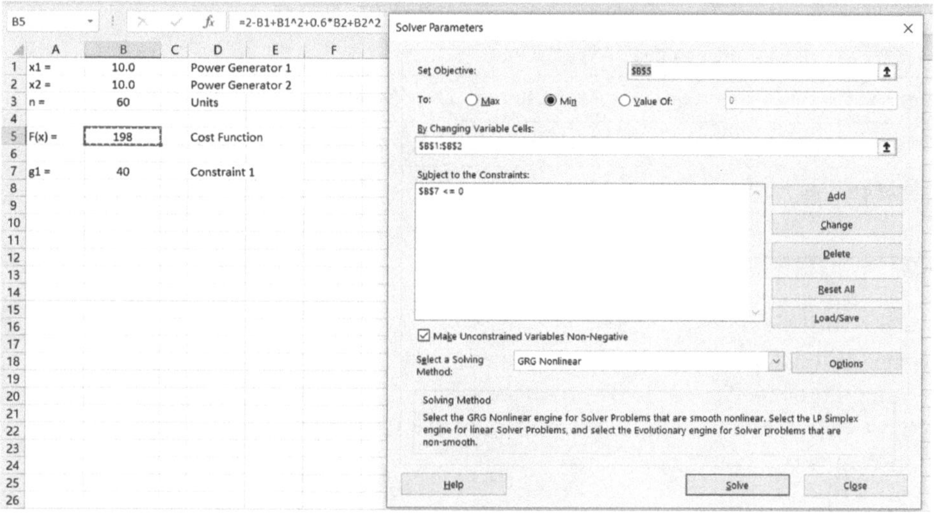

Figure 9.23: Excel Solver window of Example 3.4.

The final design obtained is shown in Figure 9.24, $\mathbf{x}^\star$ = [30.4 29.6] and $f(\mathbf{x}^\star)$ = 1790. Observe that the solution found here is the same of that shown in Section 3.4.

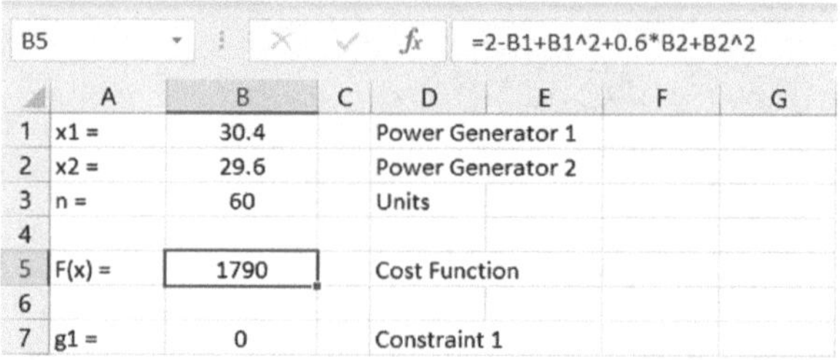

Figure 9.24: Numerical values of Example 3.4 for the final design.

9.8 Solving Example 4.2: operations research for an oil refinery

This example was solved in Section 4.2 using Simplex LP. Now it will be solved using GRG Nonlinear method. Just to remember, look at Table 9.4 where the main data of the problem is summarized. It deals with operations research for an oil refinery. The question is: how much gas of each type should be processed for maximum profit?

Table 9.4: Problem data.

Resources	Product		Available resources
	Common gas	Special gas	
Gross gas	7 m³/ton	11 m³/ton	77 m³/week
Production time	10 h/ton	8 h/ton	80 h/week
Storage	9 ton	6 ton	
Profit	$ 150/ton	$ 175/ton	

Design variables:

x_1: amount of common gas to be produced

x_2: amount of special gas to be produced

Objective function (maximize the profit): $F(\mathbf{x}) = 150x_1 + 175x_2$

Constraints:

$$g_1 = 7x_1 + 11x_2 - 77 \leq 0$$

$$g_2 = 10x_1 + 8x_2 - 80 \leq 0$$

$$g_3 = x_1 - 9 \leq 0$$

$$g_4 = x_2 - 6 \leq 0$$

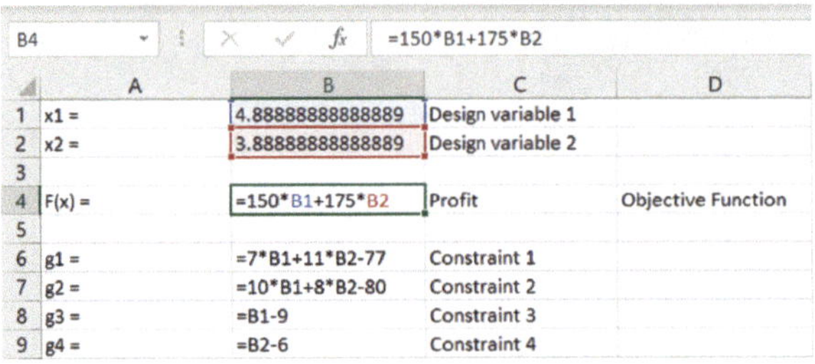

Figure 9.25: Expressions of Example 4.2.

The expressions in the Excel spreadsheet can be seen in Figure 9.25. Cells B1 and B2 are the design variables. In cell B4 the objective function is written and in cells B6 to B9 the constraints are shown.

In Figure 9.26 the initial values of the design variables and the objective and constraint functions are shown. Observe that in this design there are slacks to optimize the problem, since all constraints are negative and none is equal to zero.

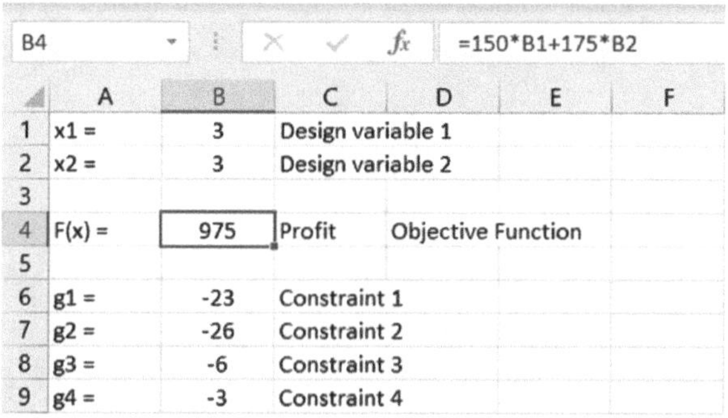

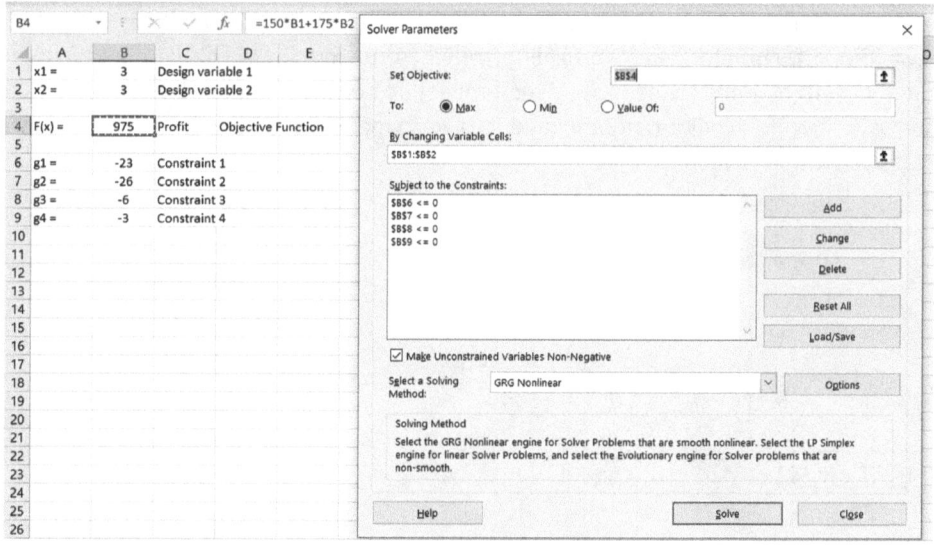

Figure 9.26: Numerical values of Example 4.2 for the initial design.

The window of the Excel Solver can be seen in Figure 9.27.

Figure 9.27: Excel Solver window of Example 4.2.

The final design obtained is shown in Figure 9.28, $\mathbf{x}^\star = [4.889 \ 3.889]$ and $f(\mathbf{x}^\star) = 1{,}413.889$. Observe that the solution found here is the same of that shown in Section 4.2. Just to remember, in Section 4.2 Simplex LP was used and here GRG Nonlinear is used.

B4			×	✓	f_x	=150*B1+175*B2	

	A	B	C	D	E	F
1	x1 =	4.888889	Design variable 1			
2	x2 =	3.888889	Design variable 2			
3						
4	F(x) =	1413.889	Profit	Objective Function		
5						
6	g1 =	-6.1E-08	Constraint 1			
7	g2 =	0	Constraint 2			
8	g3 =	-4.11111	Constraint 3			
9	g4 =	-2.11111	Constraint 4			

Figure 9.28: Numerical values of Example 4.2 for the final design.

9.9 Reliability of a commercial operation: part 1

Commercial operations must deal with several uncertainties, all of them related to costs (import, shipping, etc.) and sale price. These uncertainties can be quantified by random variables and related to a given probability density function (PDF). Figure 9.29 shows a typical PDF of a random variable. The shape of this PDF is similar to the normal distribution function.

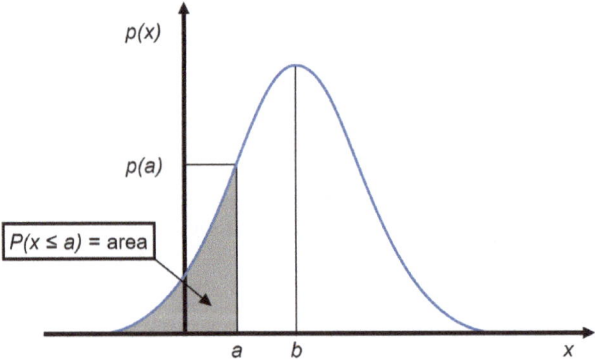

Figure 9.29: A typical probability density function.

The probability of a random variable x be less than a is the area shown in Figure 9.29. The function that represents this area is denoted cumulative distribution function (CDF) and denoted as Φ for the normal distribution (Gaussian). When a problem is dealing with several variables, each one has a PDF and a CDF. The challenge is how to compute the probability of the operation failure in this case when there are several uncertainties. That is what will be described here.

Each random variable x_i has an average (m_i) and a standard deviation (s_i) value associated. The reduced random variables can be written as

$$y_i = (x_i - m_i)/s_i$$

It is possible to compute the distance from a point $\mathbf{Y} = (y, y_2, ..., y_n)$ to the origin $\mathbf{O}$ in the reduced space as

$$\beta = \left(y_1^2 + y_2^2 + \cdots + y_n^2 \right)^{1/2}$$

The minimum value of $\beta = \beta^*$ is the reliability index. Once β is known, the probability of failure can be calculated as

$$P_f = \Phi(-\beta)$$

In Figure 9.30 the relation between β and the probability of failure (P_f) is shown.

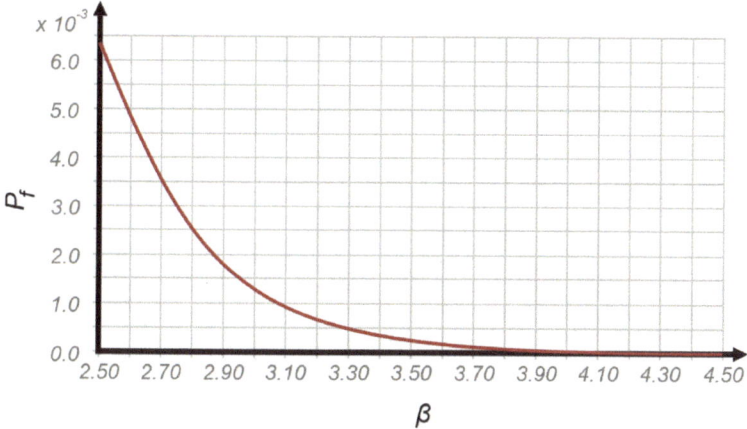

Figure 9.30: Relation between β and the probability of failure (P_f).

In this section, a commercial operation of importing a given product and selling it in a specific market will be analyzed. The following variables will be considered:

- x_1 – shipping cost ($);
- x_2 – import cost ($);
- x_3 – sell price ($).

The margin of profit, or performance function, can be written as

$$M = x_3 - x_2 - x_1$$

If the margin is positive, the seller will make a profit; if it is negative, he will suffer, and if it is equal to zero, he will have drawn.

The optimization problem (reliability problem) to determine the probability of the seller have profit can be written as

Determine **y** that minimize

$$f(\mathbf{y}) = \beta = \left(y_1^2 + y_2^2 + y_3^2\right)^{1/2}$$

subjected to

$$g_1(\mathbf{x}) = M = x_3 - x_2 - x_1 = 0$$

Note that **x** can be written as a function of **y**, as $x_i = s\, y_i + m_i$. The main data of the problem can be seen in Table 9.5. There, one can see the description of the design variables, as well as the average and the standard deviation.

Table 9.5: Problem data.

x	m ($)	s ($)	Description
x_1	20.00	10.00	Shipping Cost
x_2	200.00	30.00	Import Cost
x_3	300.00	30.00	Sell Price

B8			f_x	=(B2^2+B3^2+B4^2)^0.5	
	A	B	C	D	E	F
1	i	y	m ($)	s ($)	x ($)	Description
2	1	0	20	10	=C2+D2*B2	Shipping Cost
3	2	0	200	30	=C3+D3*B3	Import Cost
4	3	0	300	30	=C4+D4*B4	Sell Price
5						
6	M =	=E4-E3-E2	$	Sell Margin	Equality Constraint	
7						
8	β =	=(B2^2+B3^2+B4^2)^0.5		Reliability Index	Objective Function	
9						
10	Pf(b) =	=NORM.S.DIST(-B8;1)		Probability of Failure		

Figure 9.31: Expressions of Excel spreadsheet for Example 9.9.

The expressions in the Excel spreadsheet can be seen in Figure 9.31. Cells B2, B3, and B4 are the design variables. In cell B8 the objective function is written, and in cell B6 the equality constraint is written.

In Figure 9.32 the initial values of the design variables and the objective and constraint functions are shown. Note that when **y** = **0**, **x** = **m**, so the initial adopted design is the average project. Observe that the value of M, in this point, is positive equal to $80,00. It is an important information and, if this value is negative, the sign of β must be changed.

| B8 | ▼ : | × | ✓ | f_x | =(B2^2+B3^2+B4^2)^0.5 |

	A	B	C	D	E	F
1	i	y	m ($)	s ($)	x ($)	Description
2	1	0.000	20.00	10.00	20.00	Shipping Cost
3	2	0.000	200.00	30.00	200.00	Import Cost
4	3	0.000	300.00	30.00	300.00	Sell Price
5						
6	M =	80.0	$	Sell Margin	Equality Constraint	
7						
8	β =	0.00		Reliability Index	Objective Function	
9						
10	Pf(b) =	50.00%		Probability of Failure		

Figure 9.32: Numerical values of the initial design for Example 9.9.

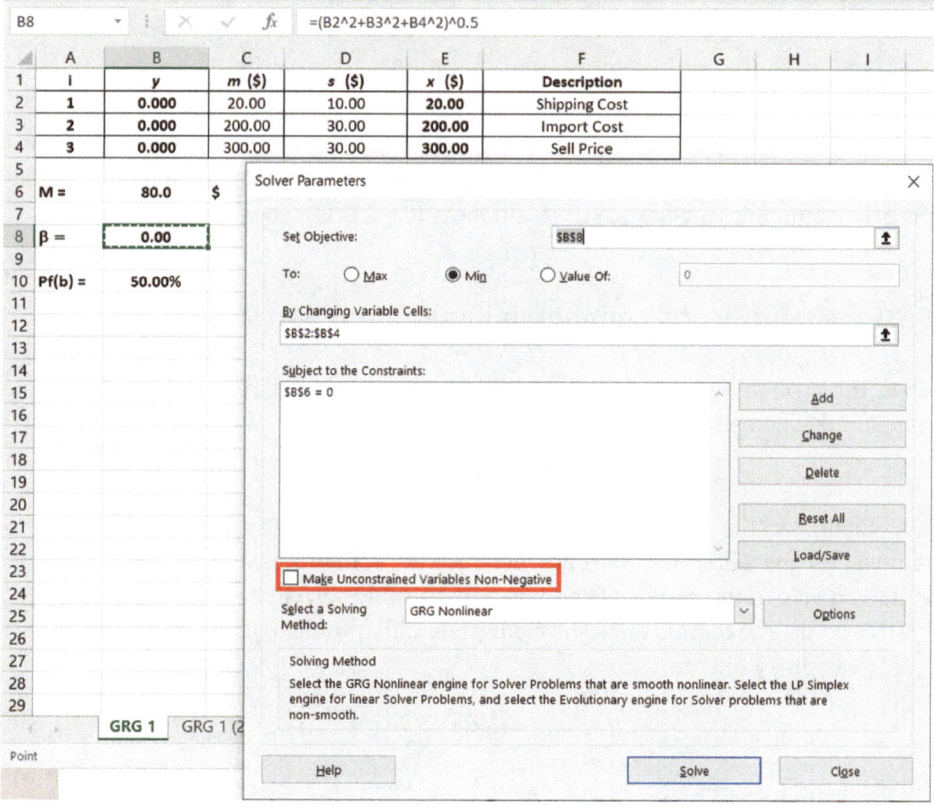

Figure 9.33: Excel Solver window of Example 9.9.

The window of the Excel Solver can be seen in Figure 9.33. Note that the box "Make Unconstrained Variables Non-Negative" must be unchecked, since the design variables can have any sign. In Figure 9.34 is shown the Excel Solver Window.

B8			f_x	=(B2^2+B3^2+B4^2)^0.5	
	A	B	C	D	E	F
1	i	y	m ($)	s ($)	x ($)	Description
2	1	0.421	20.00	10.00	24.21	Shipping Cost
3	2	1.263	200.00	30.00	237.89	Import Cost
4	3	-1.263	300.00	30.00	262.11	Sell Price
5						
6	M =	0.00	$	Sell Margin	Equality Constraint	
7						
8	β =	1.84		Reliability Index	Objective Function	
9						
10	Pf(b) =	3.32%		Probability of Failure		

Figure 9.34: Numerical values of Example 9.9 for the final design.

The final design obtained is shown in Figure 9.28, $\mathbf{y}^* = [0.421\ 1.263 - 1.263]$, with $\mathbf{x}^* = [24.21\ 237.89\ 262.11]$ and $f(\mathbf{y}^*) = \beta^* = 1.84$. The probability of the seller not achieving the desired profit is very low $P_f = 3.32\%$. It looks like this is a good business.

9.10 Reliability of a commercial operation: part 2

Now, the probability of the profits needs to be more than 20%. The performance function now is written as

$$M = 0.8x_3 - x_2 - x_1$$

Note that 0.8 in the above equation means $100\% - 20\% = 80\%$. This could represent the need of the seller to obtain at least 20% of profits to pay some fixed expenses and to survive in the market. The same data of Table 9.5 will be kept.

With this, the optimization problem (reliability problem) can now be stated as
Determine $\mathbf{y}$ that minimize

$$f(\mathbf{y}) = \beta = \left(y_1^2 + y_2^2 + y_3^2\right)^{1/2}$$

subject to

$$g_1(\mathbf{x}) = M = 0.8x_3 - x_2 - x_1 = 0$$

B8			f_x	=(B2^2+B3^2+B4^2)^0.5		

	A	B	C	D	E	F
1	i	y	m ($)	s ($)	x ($)	Description
2	1	0	20	10	=C2+D2*B2	Shipping Cost
3	2	0	200	30	=C3+D3*B3	Import Cost
4	3	0	300	30	=C4+D4*B4	Sell Price
5						
6	M =	=0.8*E4-E3-E2	$	Sell Margin	Equality Constraint	
7						
8	β =	=(B2^2+B3^2+B4^2)^0.5		Reliability Index	Objective Function	
9						
10	Pf(b) =	=NORM.S.DIST(-B8;1)		Probability of Failure		

Figure 9.35: Expressions of Excel spreadsheet for Example 9.10.

The expressions in the Excel spreadsheet can be seen in Figure 9.35. Cells B2, B3, and B4 are the design variables. In cell B8 the objective function is written, and in cell B6 the equality constraint is written. Note that now 0.8 multiplies the cell E4.

B8			f_x	=(B2^2+B3^2+B4^2)^0.5		

	A	B	C	D	E	F
1	i	y	m ($)	s ($)	x ($)	Description
2	1	0.000	20.00	10.00	20.00	Shipping Cost
3	2	0.000	200.00	30.00	200.00	Import Cost
4	3	0.000	300.00	30.00	300.00	Sell Price
5						
6	M =	20.00	$	Sell Margin	Equality Constraint	
7						
8	β =	0.00		Reliability Index	Objective Function	
9						
10	Pf(b) =	50.00%		Probability of Failure		

Figure 9.36: Numerical values of the initial design for Example 9.10.

In Figure 9.36 the initial values of the design variables and the objective and constraint functions are shown.

The window of the Excel Solver can be seen in Figure 9.37.

The final design obtained is shown in Figure 9.38 $\mathbf{y}^* = [0.127\ 0.381 - 0.305]$, with $\mathbf{x}^* = [21.27\ 211.42\ 290.86]$ and $f(\mathbf{y}^*) = \beta^* = 0.5$. The probability of the seller not achieving 20% of profits is significative $P_f = 30.72\%$. This may mean that the seller needs to be attentive to the details of his business in order not to lose money.

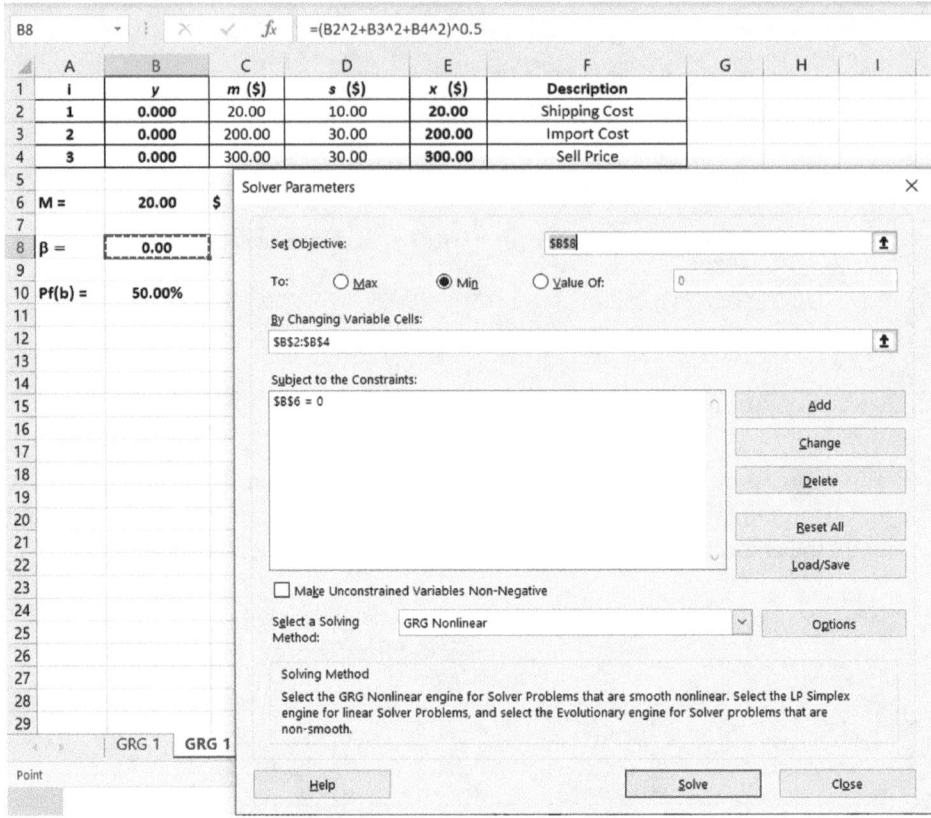

Figure 9.37: Excel Solver window of Example 9.10.

B8		f_x	=(B2^2+B3^2+B4^2)^0.5		
	A	B	C	D	E	F
1	i	y	m ($)	s ($)	x ($)	Description
2	1	0.127	20.00	10.00	21.27	Shipping Cost
3	2	0.381	200.00	30.00	211.42	Import Cost
4	3	-0.305	300.00	30.00	290.86	Sell Price
5						
6	M =	0.00	$	Sell Margin	Equality Constraint
7						
8	β =	0.50		Reliability Index	Objective Function
9						
10	Pf(b) =	30.72%		Probability of Failure

Figure 9.38: Numerical values of Example 9.10 for the final design.

9.11 Other features of the Excel Solver window

When a user clicks the button "Options" as shown in Figure 9.37, for example, it will find windows shown in Figures 9.39 and 9.40.

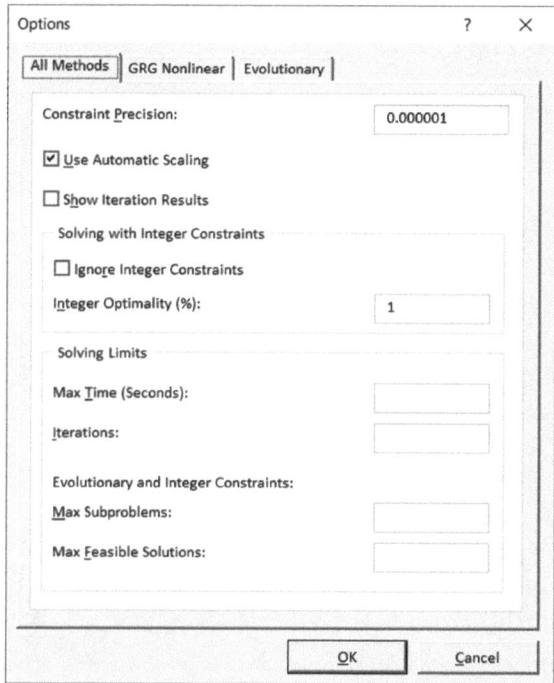

Figure 9.39: "All Methods" window of Excel Solver.

In Figure 9.39 one can see the box "Use Automatic Scaling" checked. This is important, especially when the problem presents a different order of magnitudes for the design variables. The suggestion for the other options is to consult the user manual or do specific tests on your computer. This is also true for the window in Figure 9.40.

In case of Figure 9.40 one important thing is the method to compute the derivative of the functions (objective and constraints) involved in the optimization problem: "Forward" or "Central." The first one has less precision and consumes less computational time than the second. Consequently, the "Central" presents better precision and more computational time.

In case of "Evolutionary," it was not explored in this chapter. This type of method was covered in Chapter 7 of this book.

Figure 9.40: "GRG Nonlinear" window of Excel Solver.

Appendix

MATLAB scripts

In this appendix, sample MATLAB scripts are presented to solve proposed problems of Chapter 4, on linear programming, and Chapter 6, on the use of MATLAB optimization toolbox.

A.1 Linear programming (simplex method): linearly constrained profit maximization of a toy factory production

```
%File name: linear_programming.m
%linear programming - Simplex method
% Prof. Reyolando Brasil - May 2021
%
%toy factory maximum profit
%
%main program
%design variables
%x1=number of type A toy sold
%x2=number of type B toy sold
%
clear
clc
%
a=[1 1 1 0 0;1/28 1/14 0 1 0;1/14 1/24 0 0 1]; %constraint equations
b=[16 1 1]; % resources vector
c=[-400 -600 0 0 0]; %objective function
%
%
[X,FVAL,EXITFLAG,OUTPUT]=linprog(c,[],[],a,b,zeros(size(c)),[],[])
%
```

A.2 Chapter 6: MATLAB toolbox

A.2.1 Example 5.2.1: nonlinear cable problem

```
%file name = cable_opt.m
%cable nonlinear problem
```

https://doi.org/10.1515/9783110625622-010

```
%main program
%Total Potential Energy
%Prof. Reyolando Brasil - May 2021
clear
clc
%Design variables: x(1) = u  x(2) = v
%
Prob_data(1)=200;%vertical load V, KN
Prob_data(2)=100;%horizontal load H, KN
Prob_data(3)=1;%cable cross section, cm²
Prob_data(4)=21000;%Young's Modulus, KN/cm²
Prob_data(5)=100;%La length, cm
Prob_data(6)=100;%Lb length, cm
Prob_data(7)=50;%N0 KN
options=optimset ('LargeScale','off','TolCon',1e-8,'TolX',1e-8);
x0=[10,30];
%optimization function call
[x,f,ExitFlag,Output]=fminsearch('cable_obj',x0,options,Prob_data);
disp('   u    v')
disp(x)
disp('Energia Potencial Total Mínima em KJ')
disp(f)
%
%subroutine objective funtion
%file name = cable_obj.m
%cable nonlinear problem
%objective function
%Total Potential Energy
%Prof. Reyolando Brasil - february 2021
%Design variables: x(1) = u  x(2) = v
function p=cable_obj(x,Prob_data)
%problem data
V=Prob_data(1);%vertical load V, KN
H=Prob_data(2);%horizontal load H, KN
A=Prob_data(3);%cable cross section, m²
E=Prob_data(4);%Young's Modulus,KN/m²
La=Prob_data(5);%La length, m
Lb=Prob_data(6);%Lb length, m
N0=Prob_data(7);%N0 KN
ka=E*A/La;kb=E*A/Lb;
%stretched length
LLa=sqrt(x(2)^2+(La+x(1))^2);
```

```
LLb=sqrt(x(2)^2+(Lb-x(1))^2);
%length change
da=LLa-La;db=LLb-Lb;
%axial forces
Na=ka*da;Nb=kb*db;
%strain energy
U=(N0+Na/2)*da+(N0+Nb/2)*db;
%work of external conservative forces
W=H*x(1)+V*x(2);
%Total Potential Energy
p=U-W;
```

A.2.2 Example 5.2.2: eccentrically loaded tubular column

```
%Main Program
% file name column.m
% optimization of tubular steel column with eccentric loading
% Prof. Reyolando Brasil May 2021
clear
clc
% options
options=optimset('LargeScale','off','TolCon',1e-8,'TolX',1e-8);
% limits of design variables, thickness and radius
Lb=[0.01 0.005];Ub=[1 0.2];
% initial design
x0=[1 0.2];
% call the constrained optimization routine
[x,FunVal,ExitFlag,Output]=...
   fmincon('column_objf',x0,[],[],[],[],Lb,Ub,'column_conf',options)
%
%Subroutine with the objective function
% file name column_objf.m
% objective function
% Prof. Reyolando Brasil 2021
function f=column_objf(x)
% renaming design variables
x1=x(1);x2=x(2);
% data
L=5.0; % column length (m)
rho=7850; % steel density (kg/m^3)
% objective function
```

```
A=2*pi*x1*x2;
f=A*L*rho; %column mass
%
%Subroutine with the constraint functions
% file name column_conf.m
% constraints
% Prof. Reyolando Brasil - May 2021
function [g,h]=column_conf(x)
% renaming design variables
x1=x(1);x2=x(2);
%data
P=50000; % vertical compressive load (N)
E=210e9; % elastic modulus (Pa)
L=5.0; % column length (m)
Sy=250e6; % allowable stress (Pa)
Delta=0.25; % allowable displacement at column top
% geometric characteristics
A=2*pi*x1*x2; % section area
W=pi*x1^2*x2; % bending modulus
I=pi*x1^3*x2; % moment of inertia
e=0.02*x1; % load eccentricity
%inequality constraints
g(1)=P/A*(1+e*A/W*sec(L*sqrt(P/E/I)))/Sy-1; % allowable stress
g(2)=1-pi^2*E*I/4/L^2/P; % buckling
g(3)=e*(sec(L*sqrt(P/E/I))-1)/Delta-1; % displacement at column top
g(4)=x1/x2/50-1; % Radius/thickness ratio
% equality constraints (none)
h=[];
```

A.2.3 Example 5.2.3: statically loaded redundant truss

```
%file name = tre_redun_opt.m
%redundant truss optimization
%main program
%Prof. Reyolando Brasil - 2021
clc
clear
%problem data
Prob_data(1)=1;%gravity load P, KN
Prob_data(2)=1000;%density, kg/m³
Prob_data(3)=10;%alowble stress, KN/m²
```

```
Prob_data(4)=100;%Young's Modulus, KN/m²
%options
options=optimset ('LargeScale','off','TolCon',1e-8,'TolX',1e-8);
%Lower and upper bounds of design variables
Lb=[0 0];Ub=[1 1];
%initial design
x0=[0.1 0.1];
%optimization function call
[x,FunVal,ExitFlag,Output]=...
   fmincon('tre_redun_obj',x0,[],[],[],[],Lb,Ub,'tre_hiper_con',options,
Prob_data)
%
% subroutine objective function
%file name = tre_redun_obj.m
%redundant truss optimization
%objective function
%Prof. Reyolando Brasil - 2021
function f=tre_redun_obj(x,Prob_data)
%design variables
x1=x(1);%vertical bar transverse section area, m²
x2=x(2);%diagonal bar transverse section area, m²
%material parameters
rho=Prob_data(2);%density, kg/m³
%objective function, truss total mass (kg)
f=rho*(3*x1+2*sqrt(2)*x2);
%
%subroutine constraint functions
%file name = tre_redun_con.m
%redundant truss optimization
%constraint equations
%Prof. Reyolando Brasil - 2021
function [g,h]=tre_redun_con(x,Prob_data)
%design variables
x1=x(1);%vertical bar transverse section area, m²
x2=x(2);%diagonal bar transverse section area, m²
%problem data
PP=Prob_data(1);%gravity load P, KN
Ta=Prob_data(3);%alowble stress, KN/m²
E=Prob_data(4);%Young's Modulus, KN/m²
%solution
K=E*[x1+x2*sqrt(2)/4 -x1;-x1 x1+x2*sqrt(2)/4];%stiffness matrix
P=[0;-PP];%loading vector
```

```
p=K\P;%displacements vector
%bars normal forces
N1=E*x1*(p(1)-p(2));
N4=-E*x2*p(2)/2;
N5=-E*x2*p(1)/2;
%inequality constraints
g(1)=(N1/x1)/Ta-1;
g(2)=(N4/x2)/Ta-1;
g(3)=(N5/x2)/Ta-1;
%equality constraints (none)
h=[];
```

A.2.4 Example 5.2.4: frequency optimization of a redundant truss

```
%main program
%file name = tre_freq_opt.m
%redundant truss optimization for frequency constraints
%Prof. Reyolando Brasil - May 2021
clc
clear
%problem data
Prob_data(1)=1000;%density, kg/m³
Prob_data(2)=100000;%Young's Modulus, N/m²
%options
options=optimset ('LargeScale','off','TolCon',1e-8,'TolX',1e-8);
%Lower and upper bounds of design variables
Lb=[0.01 0.01];Ub=[1 1];
%initial design
x0=[0.01 0.01];
%optimization function call
[x,FunVal,ExitFlag,Output]=...
  fmincon('tre_freq_obj',x0,[],[],[],[],Lb,Ub,'tre_freq_con',options,
Prob_data)
%
%subroutine objective function
%file name = tre_freq_obj.m
%redundant truss optimization with frequency constraints
%Prof. Reyolando Brasil - May 2021
function f=tre_freq_obj(x,Prob_data)
%design variables
x1=x(1);%vertical bar transverse section area, m²
```

```
x2=x(2);%diagonal bar transverse section area, m²
%material parameters
rho=Prob_data(1);%density, kg/m³
%objective function, truss total mass (kg)
f=rho*(3*x1+2*sqrt(2)*x2);
%
%subroutine constraint equations
%file name = tre_freq_con.m
%redundant truss optimization with frequency constraints
%Prof. Reyolando Brasil - May 2021
function [g,h]=tre_freq_con(x,Prob_data)
%design variables
x1=x(1);%vertical bar transverse section area, m²
x2=x(2);%diagonal bar transverse section area, m²
%problem data
rho=Prob_data(1);%density, kg/m³
E=Prob_data(2);%Young's Modulus, N/m²
%solution of eigenvalue problem
K=E*[x1+x2*sqrt(2)/4 -x1;-x1 x1+x2*sqrt(2)/4];%stiffness matrix
M=rho*[x1+x2*sqrt(2)/2 0;0 x1+x2*sqrt(2)/2];%mass matrix
frs=eig(K,M);%squared frequencies in rad/s
fhz=sqrt(sort(frs))/2/pi;%frequencies in Hz
%
%inequality constraints
g(1)=1-fhz(1);%f1 larger then 1 Hz
g(2)=fhz(2)-2;%f2 less then 2 Hz
%equality constraints (none)
h=[];
```

A.2.5 Example 5.2.5: thickness optimization of a rectangular steel plate simply supported under uniformly distributed loading and its own weight

```
%main program
%file name = plate_opt.m
%optimization of simply supported rectangular steel plate
%uniformly loaded and own weight
%design variable thickness x, m
%Prof. Reyolando Brasil - May 2021
clc
clear
%options
```

```
options=optimset ('LargeScale','off','TolCon',1e-8,'TolX',1e-8);
%Lower and uper bounds of plate thickness
Lb=0.01;Ub=0.1;%m
%initial design = initial thickness
x0=0.025;%m
Prob_data(1)=4;%lenght in x direction, m
Prob_data(2)=4;%lenght in y direction, m
Prob_data(3)=10000;%uniform loading, N/m²
Prob_data(4)=7850;%steel density, kg/m³
Prob_data(5)=15e7;%steel alowble stress, N/m²
Prob_data(6)=210e9;%steel Young's Modulus, N/m²
Prob_data(7)=0.3;%Poisson's ratio
%optimization function call
[x,FunVal,ExitFlag,Output]=...
   fmincon('plate_obj',x0,[],[],[],[],Lb,Ub,'plate_con',options,
Prob_data)
%
%subroutine objective function
%file name = plate_obj.m
%optimization of simply supported rectangular steel plate
%uniformly loaded and own weight
%design variable thickness x, m
%objective function
%Prof. Reyolando Brasil - May 2021
function f=plate_obj(x,Prob_data)
a=Prob_data(1);%lenght in x direction, m
b=Prob_data(2);%lenght in y direction, m
rho=Prob_data(4);%steel density, kg/m³
%objective function, the plate total mass, kg
f=rho*a*b*x;
%
%subroutine constraint functions
%file name = plate_con.m
%optimization of simply supported rectangular steel plate
%uniformly loaded and own weight
%design variable thickness x, m
%Prof. Reyolando Brasil - May 2021
function [g,h]=plate_con(x,Prob_data)
%problem data
a=Prob_data(1);%lenght in x direction, m
b=Prob_data(2);%lenght in y direction, m
q=Prob_data(3);%uniform loading, N/m²
```

```
rho=Prob_data(4);%steel density, kg/m³
Ta=Prob_data(5);%steel alowble stress, N/m²
E=Prob_data(6);%steel Young's Modulus, N/m²
nu=Prob_data(7);%Poisson's ratio
%
D=E*x^3/12/(1-nu^2);
w=x^2/6;
%
r=round(b/a,1);%b/a ratio, between 1 and 2, 11 possible values, round to first
decimal
k=10*(r-1)+1;%table position
qmp=q+10*rho*x;%gravity acceleration 10 m/s²
%
%Table 8 "Theory of Plates an Shells", Timoshenko
%
alfa=[0.00406; 0.00485; 0.00564; 0.00638; 0.00705; 0.00772; 0.00830;
0.00883; 0.00931; 0.00974; 0.01013];
%
beta=[0.0479; 0.0554; 0.0627; 0.0694; 0.0755; 0.0812; 0.0862; 0.0908;
0.0948; 0.0985; 0.1017];
%
vmax=alfa(k)*qmp*a^4/D;%maximum vertical displacement, m
%
Mx_max=beta(k)*qmp*a^2;%maximun bending moment Mx, Nm/m
%constraints
g(1)=(Mx_max/w)/Ta-1;
g(2)=400*vmax/a-1;
%
h=[];
```

A.2.6 Example 5.2.6: redundant wood planar portal frame

```
%main program
%file name = port_wood_opt.m
%optimization of redundant wood planar portal frame
%Prof. Reyolando Brasil - May 2021
clc
clear
%portal frame data
h=3;%clumn height, m
L=6;%beam span, m
```

```
b=0.075;%section b dimension, m
P=15;%design load at bem mid span, KN
%options
options=optimset ('LargeScale','off','TolCon',1e-8,'TolX',1e-8);
%Variables lower and upper bounds
Lb=[0.15 0.15];Ub=[0.4 0.4];
%initial dimensions
x0=[0.15 0.15];
%optimization function call
[x,FunVal,ExitFlag,Output]=...
  fmincon('port_wood_obj',x0,[],[],[],[],Lb,Ub,'port_wood_con',options,
h,L,b,P)
%
%subroutine objective function
%file name = port_wood_obj.m
%optimization of redundant wood planar portal frame
%Prof. Reyolando Brasil - May 2021
function f=port_wood_obj(x,h,L,b,P)
%design variables
x1=x(1);%column section d dimension, m
x2=x(2);%beam section d dimension, m
%wood parameters
rho=1000;%density, kg/m³
%objective function, portal frame total mass
f=rho*(2*h*b*x1+L*b*x2);
%
%subroutine constraint equations
%file name = port_wood_con.m
%optimization of redundant wood planar portal frame
%Prof. Reyolando Brasil - May 2021
function [g,h]=port_wood_con(x,h,L,b,P)
%
%design variables
x1=x(1);%column section d dimension, m
x2=x(2);%beam section d dimension, m
Ac=b*x1;%column section area, m²
Ab=b*x2;%beam section area, m²
Wc=b*x1^2/6;%column section flexural modulus, m³
Wb=b*x2^2/6;%beam section flexural modulus, m³
Ic=b*x1^3/12;%column section moment of inertia, m^4
Ib=b*x2^3/12;%beam section moment of inertia, m^4
%wood parameters
```

```
E=15e6;%efective Young's Modulus, KN/m²
fcd=20e3;%design resistance, KN/m²
%axial forces and bending moments
X=(P*L^2*h/8/Ib)/(2*h^3/3/Ic+L*h^2/Ib);
Nc=P/2;%KN
Mc=X*h;%KNm
%
FE=pi^2*E*Ic/h^2;%Column Euler's buckling load
e=(Mc/Nc+h/300)*(FE/(FE-Nc));
Mc=Nc*e;
%
Nb=X;%KN;
Mb=P*L/4-X*h;%KNm
%
FE=pi^2*E*Ib/L^2;%Beam Euler's buckling load
e=(Mb/Nb+L/300)*(FE/(FE-Nb));
Mb=Nb*e;
%
%inequlitiy constraints
g(1)=(Nc/Ac+Mc/Wc)/fcd-1;
g(2)=(Nb/Ab+Mb/Wb)/fcd-1;
%equality constraints (none)
h=[];
```

References

Arora JS. Optimization of structures subjected to dynamic loads, structural dynamic systems, computational techniques and optimization. Optimization Techniques. 1999; 1–72.

Arora JS. Introduction to Optimum Design. 3rd Ed. Oxford: Academic Press; 2012.

Arora JS, Chahande AI, Paeng JK. Multiplier methods for engineering optimization. International Journal for Numerical Methods in Engineering. 1991; 32: 1485–1525.

Arora JS, Huang MW, Hsieh CC. Methods for optimization of nonlinear problems with discrete variables: a review. Structural Optimization. 1994; 90: 69–85.

Artelys. *Artelys Knitro User's Manual*. Available in <https://www.artelys.com/docs/knitro/3_referen ceManual/knitromatlabReference.html#knitro-nlp>. Acessed in Februray 10, 2021.

Atkinson K. Elementary Numerical Analysis. Hoboken: Wiley; 1993.

Bendsoe MP, Sigmund O. Topology optimization: Theory, Methods and Application. Berlin: Springer-Verlag; 2003.

Brasil RMLRF, Silva MA. Introdução à Dinâmica das Estruturas para a Engenharia Civil. 2nd Ed. São Paulo: Ed. Edgard Blucher; 2015 (in Portuguese).

Brasil RMLRF, Balthazar JM, Gois W. Métodos Numéricos e Computacionais na Prática de Engenharias e Ciências. São Paulo: Ed. Edgard Blucher; 2015 (in Portuguese).

Brasil RMLRF. Bases da Mecânica dos Sólidos Elásticos com Elementos Finitos. Santo André: Editora UFABC; 2017 (in Portuguese).

Burnett DS. Finite Element Analysis. New York: Addison Wesley; 1987.

Chahande AI, Arora JS. Optimization of large structures subjected to dynamic loads with the multiplier method. International Journal for Numerical Methods in Engineering, 1994; 37: 413–430.

Chapra SC, Canale RP. Numerical Methods for Engineers. 5th Ed. New York: McGraw-Hill; 2006.

Ditlevsen O, Madsen HO. Structural Reliability Methods. Hoboken: Wiley; 2005.

Fletcher R. Practical Methods of Optimization. Hoboken: Wiley; 1985.

Haug EJ, Arora, JS. Applied Optimal Design. Hoboken: Wiley; 1979.

Jin G. Monte Carlo finite element method of structure reliability analysis. Reliability Engineering & Systems Safety. 1993;40: 77–83

Johnson LW, Riess RD. Numerical Analysis. New York: Addison Wesley; 1982.

Kikuchi N. Finite Element Methods in Mechanics. Cambridge: University Press; 1986.

Kocer F. Arora JS. Optimal Design of Nonlinear Structures Subjected to Dynamic Loads with Application to Transmission Line Structures, University of Iowa, Optimal Design Laboratory, Technical Report No. ODL-910.02; 1999.

López CP. MATLAB Optimization Techniques. New York: Springer; 2014.

Melchers RE, Beck AT. Structural Reliability Analysis and Prediction. 3rd Ed. Wiley; 2018.

Nowak AS, Collins KR. Reliability of Structures. 2nd Ed. Boca Raton: CRC Press; 2012.

Piskunov N. Integral and Differential Calculus. Moscow: Ed. Mir; 1979.

Polak E. Computational Methods in Optimization – A Unified Approach. Oxford: Academic Press; 1971.

Rao SS. Engineering Optimization, Theory and Practice. Hoboken: Wiley; 2009.

Rausand M, Hayland A. System Reliability Theory: Models, Statistical Methods, and Applications. 2nd Ed. Hoboken: Wiley Series in Probability and Statistics; 2003.

Rocha DC, Silva MA, Brasil RMLRF. Otimização de Torres Metálicas Para Suporte de Geradores Eólicos. In: Anais CILAMCE. 2016 (in Portuguese).

Silva MA. Aplicação do Lagrangeano Aumentado em Otimização Estrutural com Restrições Dinâmicas [Master's Thesis]. São Paulo: Escola Politécnica, Universidade de São Paulo; 2000 (in Portuguese).

https://doi.org/10.1515/9783110625622-011

Silva MA. Sobre a Otimização de Estruturas Submetidas a Carregamentos Dinâmicos [dissertation]. São Paulo: Escola Politécnica, Universidade de São Paulo; 2000 (in Portuguese).

Silva MA, Brasil RMLRF. O Cálculo Simultâneo do Equilíbrio e da Confiabilidade de Seções de Concreto Armado Utilizando-se Técnicas de Otimização. In: Anais 58CBC IBRACON. 2016 (in Portuguese).

Suresh K. Design Optimization Using Matlab and SolidWorks. Madison: University of Wiscosin; 2019.

Index

https://doi.org/10.1515/9783110625622-012